Portraits of the Bison

An Illustrated Guide to Bison Society

Portraits of the Bison

An Illustrated Guide to Bison Society

Wes Olson

Photography by Johane Janelle
Foreword by Clarence Tillenius

Published by

The University of Alberta Press
Ring House 2
Edmonton, Alberta, Canada T6G 2E1

ISBN 0-88864-432-9

Library and Archives Canada Cataloguing in Publication Data

Olson, Wes, 1954–
Portraits of the bison : an illustrated guide to bison society / Wes Olson ; with photography by Johane Janelle ; foreword by Clarence Tillenius.

Includes bibliographical references.
ISBN 0-88864-432-9

1. American bison—Behavior. 2. Social behavior in animals. 3. Bison attacks—Prevention. I. Janelle, Johane, 1959– II. Title.

QL737.U53O57 2005 599.64'3156 C2005-903361-4

Printed and bound in Canada by Friesens, Altona, Manitoba.
First edition, first printing, 2005

The publisher gratefully acknowledges Dr. Margaret Hess's generous donation in establishing a revolving fund for the University of Alberta Press that has helped make possible the publication of this book.

Photo Credits

All photographs by Johane Janelle, except those by Wes Olson on: half title page—left and right, VI, VIII, X, 2, 6, 8 bottom, 12, 13 top right, 14, 16, 20, 22, 26, 27, 29, 32, 35, 36, 37, 38, 39, 42, 43, 44 bottom, 48 top left, 53 top right and bottom right, 72, 73, 76, 80, 81, 84, 85, 86, 88, 90, 96 top left, bottom left and bottom right, 97, 98 top right and bottom right, 99, 100 bottom left and right, 101, 102, 103, 108 right.

Historic photographs on pages XII–1 (PAA A.3994), 10 (PAA A.3994), 9 (PAA P.454), and 13 (PAA A.4725), top left, are used with permission of the Provincial Archives of Alberta.

The University of Alberta Press gratefully acknowledges the support received for its publishing program from The Canada Council for the Arts. The University of Alberta Press also gratefully acknowledges the financial support of the Government of Canada through the Book Publishing Industry Development Program (BPDIP) and from the Alberta Foundation for the Arts for our publishing activities.

Contents

A late spring snow storm in Elk Island National Park, Alberta.

Foreword

More than any other animal, the buffalo has become the icon of the Old West. From Mexico to Alaska, the opportunity to see again on the prairies this remnant of the vanished millions draws ever-increasing thousands of tourists simply to catch a close-up view of bison.

Many, many books have been written about this great beast, but no book exactly like this one, written and drawn by author Wes Olson, has yet been placed before the public.

Here is a man born and raised in the foothills of Alberta, the very home range of the buffalo. For the past twenty years as a warden of Elk Island National Park, his whole and exclusive drive has been to study and be with the buffalo (*bison* to the scientist), to learn its habits, how these animals interact with each other from birth to old age, how they respond to new situations, and how these mighty and sometimes unpredictable creatures react to people who more and more want to come closer and closer to them.

Unpredictable? Yes. Wes points out that every year more people are injured by bison than by bears, a fact surprising to most park visitors.

In this book, Wes sets out a considerable number of guiding cautions that should be studied, memorized, and taken to heart by anyone planning to visit a buffalo herd on its own turf anywhere in North America.

My own acquaintance with the buffalo began some seventy years ago and my study of them has gone on since that time, and so I can recommend without reserve this guide to bison behaviour. But it is much more than that. Here are drawings and diagrams and descriptive analyses—from newborn calf to aged bull—of the plains bison of North America.

For more than thirty years Wes Olson has worked with wildlife, a natural career for someone whose childhood was spent in regular outings to the woods and the mountains, there to be among bighorn sheep, moose, elk, deer, and bears.

After a two-year stint at Lethbridge Community College in wildlife research, he headed to the Yukon Territory, where he worked assisting wildlife biologists in all manner of research, working with every ungulate species found in the North as well as with black bears and grizzly bears, small mammals, and waterfowl, and even doing a little undercover law enforcement.

Then it was back to Alberta, this time surveying moose and grizzly bear populations and the impact of oil exploration and development on them. His work included two summers of live trapping and fitting radio collars on black bears and grizzly bears, with the winters spent in aerial surveys for moose and elk.

A stint with the Banff School of Fine Arts provided a break from this, which resulted in his falling in love with Banff National Park and also in his first meeting with the bison living there. From there he moved on to Waterton Lakes National Park for a season, then back to Banff, and in the spring of 1984 to Elk Island National Park, where his whole interest and dedication ever since has been to the buffalo.

This book is the fruit of that dedication, and as a long-time bison enthusiast I can unreservedly recommend it to anyone who even contemplates going near a buffalo.

Clarence Tillenius

Acknowledgements

WHILE THE MAJORITY of the information in this book is a direct result of two decades of observing bison in Elk Island National Park, as well as in numerous other national and state parks, sanctuaries, and reserves, several authors deserve special acknowledgement. The published works of Joel Berger, Dale Lott, Wendy Green, Petr Komers, and Stan van Zyll de Jong helped me to understand bison social structure and to place what I was observing in the field into a form that I could then portray through the text, drawings, and photographs within this book. Jill Fallis and Bonny Yates provided invaluable editing of my poor grammar and punctuation. Wendy Johnson drew the map on page 4 and made additions to the NASA photograph on page 5. Mary Meagher and Delaney Boyd-Burton reviewed the draft manuscript and provided invaluable advice that helped make the manuscript more readable and accurate. Thank you.

No acknowledgment would be complete without thanking my wife, Johane. Without her support and encouragement through the many years this has taken to create, the book would have died a lonely death on my drafting table. Instead it has come to life, and hopefully it will enhance the lives of those who read it.

The University of Alberta Press would like to thank Dr. Margaret Hess for her generous donation in establishing a revolving fund to assist our publications program. *Portraits of the Bison* is the first book to benefit from this fund.

Once bison trails such as this traversed much of North America in a myriad of interconnecting highways. Today these bison trails have been replaced with paved roads. Via these new trails, the modern-day explorer can travel to parks and wildernesses for the opportunity to tramp along paths trod deep by passing bison.

Introduction

BUFFALO! The word alone invokes instant feelings of nostalgia, romance, and times more adventurous and daring. It brings to mind the era of Buffalo Bill Cody, Kit Carson, and lesser-known yet just as pivotal characters like Samuel Walking Coyote, Michel Pablo, Charles Allard, Charles Goodnight, and C.J. "Buffalo" Jones, to name but a few. It calls forth images of wagon trains, hide hunters, Native peoples on horseback, tipis, and the eventual guilt and remorse of a slaughter unprecedented.

Try to imagine tens of millions of *anything*, let alone that many animals the size of a buffalo! If you were able to line up that number of buffalo from nose to tail, they would form a line long enough to go around the earth's equator twice! Imagine that many—it takes your breath away. And we almost lost them all. Once the slaughter was over and the first conservation efforts began, fewer than one hundred free-roaming plains bison remained in the entire world. If you were to line up those bison from nose to tail, there would barely be enough to go a dozen city blocks.

Often when I am sitting quietly along the edge of a stream watching a bull drink, his beard floating downstream, or resting with my back against a tree trunk along the outer edge of a forest, or sitting in the centre of a meadow surrounded by these great, massive, furry beasts, I marvel at my luck and my good fortune, to be able to share that space and time with them. As I sit quietly, I hear the soft enquiring grunts of the cows as they talk to their calves, and the gentle grunt in return as the calf answers, "Here I am." I listen to the sound of their breathing, of their teeth ripping grass from the ground, of their hooves clacking and their tails swishing. I watch the antics of brown-headed cowbirds, a species which coevolved with bison, as they pick lice and ticks from the backs of these passing grocery stores. I watch as an obviously nursing coyote fades, ghostlike, in and out of sight among the herd, using the bison as cover as she deftly hunts for ground squirrels and other small mammals to feed her pups in a nearby den. And I smell them, and I marvel at my ability to do so. How often can you say, "I smelled a buffalo today"?

Every time I experience these little things, I treasure them and I give silent thanks to the people who had the foresight to protect the last remnants of this magnificent species. For had they not done so, I would not have the opportunity to walk along trails created by the hooves of passing bison; I would not be able to marvel at the effect their manure patties have on the entire surrounding ecosystem; I would not be able to sit with pencil and paintbrush in hand, creating the images that I share with you in this book.

Many people have asked me: Are they bison or buffalo? Technically, there are no buffalo in North America. True buffalo live on the Asian or African continents, while North America has both the plains bison (or prairie bison) and the wood bison of northern Canada. There exists a third type of bison, the European bison or wisent, found primarily in eastern Poland. While I use *bison* throughout the text, it is only by being careful, since in informal situations I often refer to these animals as buffalo. The name *buffalo* has a long history of use, as well as, to me, a romantic ring of tradition attached to it. So refer to them however you wish: bison or buffalo.

Several people have asked me why I wanted to produce a book about bison. There are many reasons, but they boil down to two main ones. The first is frustration. Over the years, I have travelled through many places that have free-roaming bison populations. In almost every one, I have seen people placing their lives, and the lives of their loved ones, in danger by approaching too close to bison. In Yellowstone a few years ago, my wife, Johane, and I were watching a group of rutting bison graze peacefully along the road through Hayden Valley. A car from New Jersey pulled up and out leaped two strapping young men. One of them trotted out into the meadow, then turned his back on a large bull so that his friend could take his picture. Within seconds, the aspiring model was charged by the bull and forced to run (laughing) back to his car. The fellow was lucky he was not gored.

A few minutes later, a large motorhome from New York stopped and out stepped a man with his young son. The father instructed his son to back up toward the bull while he filmed him with a video camera. If an adult knowingly places his life in jeopardy, then he must accept the consequences, but when a child is forced into danger by his father, we can not sit back and watch. So Johane yelled over to the father that this bull had just charged some other people. The man turned toward us, just long enough to send over a particularly withering glare, then instructed his son to back even closer to the bull. For some reason those two were lucky that day, or else the bull recognized that the son posed no threat to him. It is unfortunate that this family drove away from the situation thinking that their actions were safe and believing that we were wrong in warning them of the dangers associated with bison. The next time this father places his son's life at risk, they may not be so lucky.

I hope that after reading this book and examining the photos and drawings provided, visitors to these beautiful places will be a bit more respectful of the animals.

The second reason for producing this book is to give readers the opportunity to learn a little about how complex bison societies are. I clearly recall the first time I saw a large group of bison in Elk Island National Park, Alberta. Travelling through the park, I was forced to a stop while what seemed like several hundred bison slowly passed in front of me. I remember being fascinated, but unable to distinguish the males from the females. Over the next couple of decades, I made a point of educating myself, and the drawings, paintings, and photographs within this book are the result of that education.

Most of information contained here is derived from my observations at Elk Island National Park. Just as bison themselves are individuals, variations exist among bison populations, caused by regional differences in habitat, genetic background, and the active manipulation of the population by park managers. Consequently, depending upon where you are, you may encounter slight discrepancies between the size, shape, and behaviour of the bison you see and the information contained here.

If you are lucky, you too will have the privilege of witnessing the birth of calf, of watching two magnificent bulls fight a glorious battle, or of simply being able to say, "I saw a buffalo today."

1

The Bison Saga

FROM NEAR EXTINCTION TO SALVATION

A massive bull beds quietly with a group of cows and calves in Elk Island National Park. Peaceful scenes like this are common in many of the North America's bison refuges, but should a rival bull enter the picture, they can quickly erupt into bedlam.

TENS OF MILLIONS of plains bison roamed the prairies of North America during the mid-1800s. They populated the continent from the southern shores of the Great Lakes to the Allegheny Mountains, south to the Carolinas, and westward through southern Texas and into northern Mexico. These mighty beasts extended north well into central Alberta, Saskatchewan, and Manitoba, where the vast expanse of short- and tall-grass prairie, wooded valleys, and mountain slopes saw an everchanging kaleidoscope of movement as the great herds travelled across the landscape.

Herds of a hundred bison, of several thousand, of fifty or fewer, were constantly on the move, searching for a new patch of grass, a drink of fresh water, or an exposed, windy hilltop that provided refuge from the swarms of horseflies, mosquitoes, and other biting insects that constantly plagued them. Once settled, the bison would lay in all their somnolent glory and placidly, monotonously chew their cuds. These herds, in their movements across the land, would encounter other similar groups and merge, stay together for a while, then split up into new amalgamations, picking up a few new members here and losing a few there.

Modern-day cattle ranchers employ a pattern of rotational grazing to ensure that pasture lands do not become overgrazed. If the herd is kept in one place for too long, the available forage is consumed and the animals must be moved to fresh pastures. The bison of North America naturally practised this art, and while they may have temporarily devastated

This map illustrates the historic maximum distribution of plains bison and wood bison, and the approximate zone of range overlap in North America.

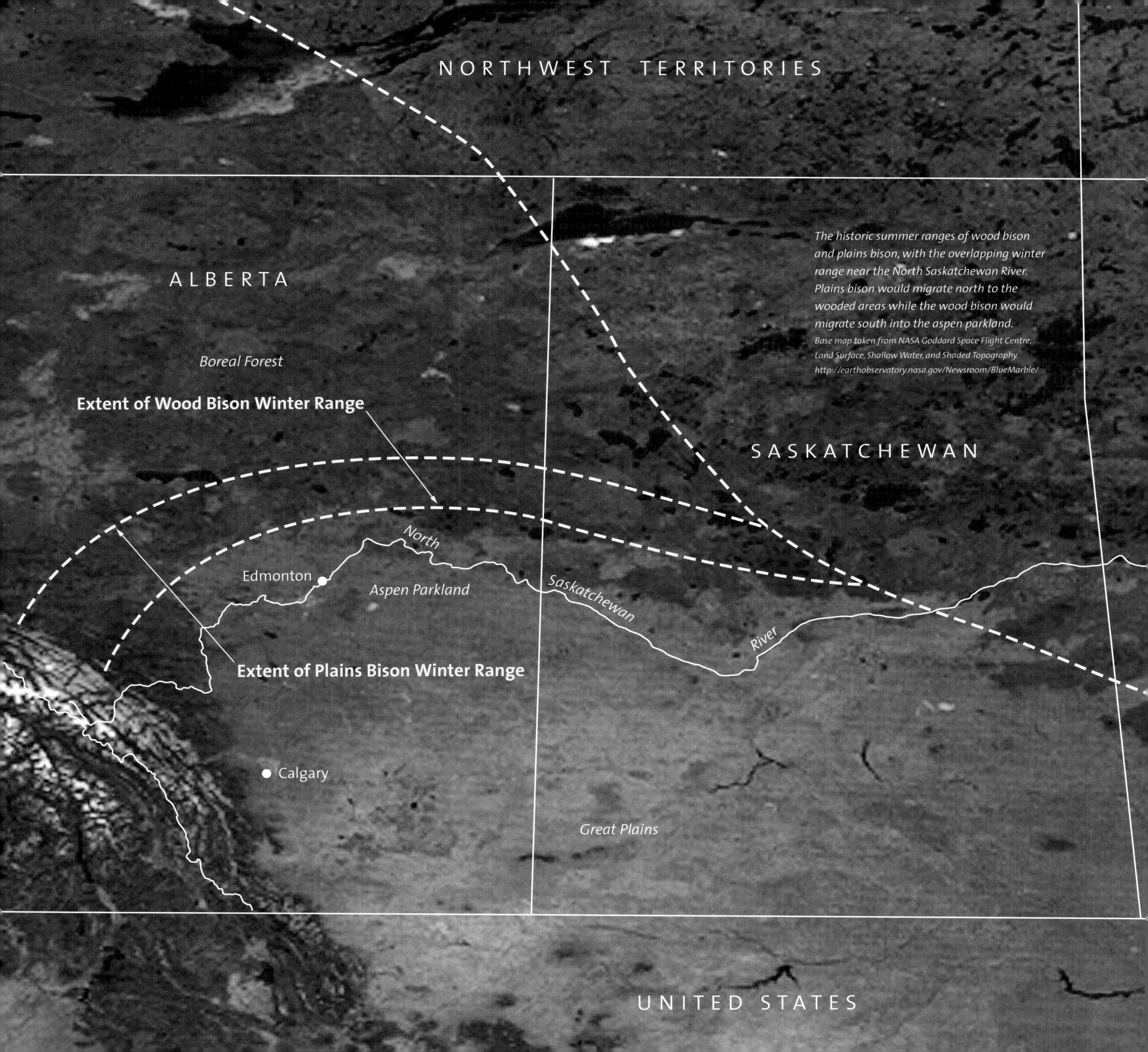

The historic summer ranges of wood bison and plains bison, with the overlapping winter range near the North Saskatchewan River. Plains bison would migrate north to the wooded areas while the wood bison would migrate south into the aspen parkland.

Base map taken from NASA Goddard Space Flight Centre, Land Surface, Shallow Water, and Shaded Topography. http://earthobservatory.nasa.gov/Newsroom/BlueMarble/

entire grasslands with their heavy use, this was part of natural cycle of intense use followed by extended rest. The herd would move on and possibly not return to the area for many years, thus allowing time for the considerable quantity of manure left behind to be incorporated into the soil, for the plants to rest, and the seeds masticated by grinding teeth time to germinate. With the bison's removal of the tall thatch of vegetation, herds of elk, pronghorn antelope, mule deer, and white-tailed deer would in turn take advantage of the greener pastures left behind. Birds of many species also benefited. Meadowlarks nested in the tall clumps of grass that surrounded buffalo patties; sage grouse and sharp-tailed grouse performed their breeding dances in the wallows left behind by passing bison; huge flocks of passenger pigeons thrived in an environment created and maintained by bison.

Small mammals also reaped the rewards of an association with bison. Black-tailed prairie dogs and Richardson's ground squirrels have a complex relationship with bison that has lasted for thousands of years. These species require a surrounding landscape of short grass that enables them to stand tall and see approaching predators, both avian and terrestrial. The black-footed ferret roamed these towns and villages, searching for the less wary squirrels. And the bison in turn benefited by using the mounds created by the ground squirrels as fresh material for their daily dustbaths.

Following these herds, and totally dependent upon them, was a host of other species. Wolves warily circled the herd's perimeter, always searching for the lame, the weak, the less vigilant. Their leftovers were fought over by the scavengers that followed behind: swift foxes, coyotes, vultures, eagles, woodpeckers, and other small birds. Mixed in with the dinner guests were a host of insects that also used the decaying carcass as a food source. By the time the last diner had left the dinner table, only bleached and whitened bones remained, and these too would be gnawed upon by mice and other rodents for the nutrients they contained.

Native peoples from many different nations and tribes in Canada and the United States depended entirely upon bison as a source of food, clothing, shelter, water containers, toys, and a host of other items. This

Pronghorn antelope at the National Bison Range, Montana (above), *and Richardson's ground squirrels in Elk Island National Park, Alberta* (below), *share the habitat and benefits associated with bison.*

Stripped clean of flesh and sinew, the skeletons of these bison at a buffalo pound near Lloydminster, Alberta, gradually returned to the soil, passing their life energy on to a host of other plant, insect, and animal species. (PAA P.454)

was the pattern of life across North America for many decades. Though not necessarily a peaceful and tranquil period—with intertribal warfare combined with the stresses associated with survival in a harsh and demanding landscape—it was a dynamic system with wildlife populations increasing and decreasing with the arrival of severe winters, major floods, or other natural catastrophes. The fate of people who relied upon these animals was dictated in part by the health of the herds.

With the arrival of Europeans during the 1700s and in particular the 1800s, a dramatic and permanent change gradually spread across the land. The demand for land for ranching and farming, the discovery of gold in California, and the constant, ever-westward movement of people soon created a land where bison were unwelcome. The extirpation of the species began in the east and quickly gained momentum as colonization spread. Numerous accounts of the demise of this magnificent animal exist, and it is beyond the scope of this book to recount them all. (For a partial list, see *Further Reading* at the back of this book.) Suffice it to say that over a period of about forty years, the once-huge herds were reduced to scattered, isolated, and very small groups.

By 1888, only an estimated eight bison remained in the wild in Canada, with another eighty-five in the United States. It is hard to

Although the plains of central North America were perceived by many people to be a bleak and barren wasteland, these huge grasslands provided everything the bison required. For those people who occupied the region, the bison in turn provided everything they needed to thrive in an otherwise harsh and inhospitable environment. (PAA A.3994)

imagine the speed with which humans decimated those awesome herds, and it is hard to imagine what the world would be like today without bison in it. It was difficult for the settlers, as they crossed the Great Plains and made their way west, to exist without the dung of the bison. In a land of endless grass, with no forests from which to harvest wood for cooking and heat, the only fuel available was dried buffalo dung. This fuel was lightweight, easily carried on a stick or ramrod, or in the aprons of women, and it was found in abundance wherever bison had been. Wagons of all types carried a sheet of canvas slung underneath. Upon there, beneath the floor of the wagon, the buffalo chips were kept removed from the necessities of daily life, yet close at hand and protected from the elements. It was essential to the settlers' survival that their fuel stayed dry, because while it burned very hot and clean when dry, it would not even smoulder when wet.

In addition to the plains bison that occupied most of central North America, another subspecies of bison existed farther north. Across the wilderness that was eventually to become northern British Columbia, Alberta, Saskatchewan, and the southern Northwest Territories, most of the Yukon, and part of Alaska lived a race of bison known today as the wood bison. These were a different-looking bison, taller at the hump and larger in body by up to a third than the plains bison, with different pelage (hair coat) characteristics. They lived in a much harsher and more demanding environment than their southern cousins, and their body shape and traits evolved from the stresses that the environment placed upon them. Wood bison probably never formed the huge rutting herds for which the plains bison were so famous. The habitat simply could not support herds of that size.

The northern ranges consisted of large and small meadows, streams and river valleys, muskegs, and swamps interspersed within a landscape known today as the boreal forest, and farther south into the aspen parkland. The southern limit of the wood bison was about 100 kilometres or

60 miles north of the North Saskatchewan River in Alberta; the northern limit of the plains bison covered roughly the same area. This implies that there was a zone of overlap between the two subspecies, and while this is correct, it was an overlap of winter ranges, not breeding ranges. Probably a few plains bison and wood bison did crossbreed, but their genetic influence would have been lost in the sea of bison around them.

Wood bison also suffered at the hand of humans, and by the end of the nineteenth century, only an estimated three hundred remained. The establishment of Wood Buffalo National Park, straddling the border of Alberta and the Northwest Territories, saw the beginning of this species' recovery. Under the protection of the park, the herd increased to about fifteen hundred by the early 1920s. Then came one of the worst decisions in the history of Canadian wildlife management.

Due in part to the Depression of the 1920s and poor fiscal management on the part of park staff, and in part because the near extinction of the species was still fresh in people's minds, the government of Canada decided to translocate 6,673 plains bison from Buffalo National Park, in Alberta, north to the wilds of Wood Buffalo National Park. The assumption at the time was that the region was so vast and the introduction site so far from the resident wood bison that the two types would not encounter each other. They did, though, and by the late 1930s the last wood bison was thought to have been lost due to hybridization with the introduced plains bison.

Even worse than the translocation of plains bison genes was the translocation of the cattle diseases bovine brucellosis and tuberculosis. The plains bison from Buffalo National Park carried these diseases with them, and within a very short while had infested the region.

In the late 1950s a remote, isolated herd of wood bison was discovered in the Nyarling River region of Wood Buffalo National Park. Given their isolation and the fact they appeared, phenotypically, to be pure wood bison, they were captured. From this small herd sprang the wood bison herds in Elk Island National Park, Alberta, and the Mackenzie Bison Sanctuary, Northwest Territories.

Some evidence suggests that a third race, or a geographic variant, of bison existed through to the early 1800s: the mountain bison.

Whether forested habitats, open grassland, or sedge meadows, the range of the wood bison covers a bounteous land of plenty during the summer. In winter, however, temperatures often fall below –40° for extended periods of time, and when combined with deep snows, the north can be a hard, unforgiving land.

According to the first explorers to reach the Rocky Mountains, this race was larger, darker, and more reclusive than the plains bison they were used to seeing. The mountain bison reportedly lived in small herds in the forested higher-altitude regions of such states as Wyoming, Washington, Idaho, and Montana. Possibly this mountain race was the original hybrid, a cross between the plains bison that moved into the eastern slopes of the Rocky Mountains and the wood bison that moved south and west into the same area. Certainly the descriptions by early explorers match the description of the modern wood bison, and the bison residing in Yellowstone National Park today bear some of the same physical traits as the northern subspecies. It is widely accepted, though, that as a distinct race (or geographic variant) they were extirpated from most of their range by about 1840.

The bison in Yellowstone today are the descendents of the only herd in North America that was founded from animals endemic to the area (though some were added later). All other plains bison herds descend from bison caught and then moved to another area. The traits of these founding animals may still be seen in the Yellowstone herd.

Whatever their source, and regardless of their type, race, or subspecies, the bison in North America today owe their very lives to the foresight and fortitude of the early conservationists and ranchers. Following in the footsteps of the first bison conservationists at the end of the nineteenth century are the bison ranchers of the twenty-first century. The century between saw the bison of North America—throughout Canada and the United States—experience a tremendous population explosion. From the slim remnant that remained at the start of the 1900s, by 2003 more than 300,000 plains bison existed on private lands and another 20,000 in public herds. (See Appendix 2 for a current listing of most public herds in North America.)

One hundred years after the Pablo-Allard herd was captured and relocated in Montana's first bison roundup, the process of gathering together and caring for bison continues. Today this scene plays itself out wherever conservation herds are maintained—on private ranches, family farms, wildlife refuges, and government reserves within the confines of national, state, and provincial parks. Photo of the annual roundup at Custer State Park, South Dakota (left).

Specially constructed barges heavily laden with their cargo of plains bison from Buffalo National Park begin their long journey north up the Athabasca River in 1925 (PAA A.4725) (top left). *A wood bison cow and bull in the Peace-Athabasca Delta region of Wood Buffalo National Park* (top right).

The original home of the wood bison, Canada's Wood Buffalo National Park, is a vast region of boreal forest with a complex and ever-changing mosaic of spruce, aspen, and birch forests interspersed with meadows, sedge lands, and countless lakes and ponds. Wood bison evolved in this remote land, and later the Canadian government transplanted over 6,000 plains bison into the park.

2

People and Bison

SAFETY AND AWARENESS

Beyond This Place, Thar Be Bufflers!

Back in the days of superstition and the belief that the world was flat, the maps of the time had a warning printed near the edge of the known world, which said *Beyond this place, thar be dragons!* This was meant to serve as a warning that there were dangerous things out there in the unknown and those brave enough to enter should tread warily. Well, there are no dragons here, but there are bison, and like the dragons of old, they are to be respected.

The information contained within this book will make your viewing of bison more interesting and enjoyable. Know, however, that while bison are not normally dangerous or aggressive, many people have been injured as a result of being lulled into a false sense of security by the bison's apparently docile nature. So before you go wandering off in pursuit of an award-winning photograph of your loved one standing next to a cranky old bull, read the following sections. Then enjoy your adventure with safety in mind.

How Close Is Too Close?

How close a person and a bison should be is a critical question that must be answered both for your safety and for the safety of the bison. To be safe, never approach a bison closer than about 75 to 100 metres or 250–330 feet. Occasionly—while walking through a forest, for example—you may inadvertently come across a herd grazing or resting along your travel route, and then you must deal with a situation where you are closer than 100 metres or 330 feet. It may seem to you, as you approach a lone bull or even a group of quietly grazing bison, that your presence is not bothering the animals because they continue to graze without a noticeable change in their behaviour. What you cannot see, however, is the increase in their heart rates and the increase in their stress level. Even though the bison outwardly appear to be fine, the closer you get, the higher their stress level becomes. They can tolerate your approach only for a limited length of time before they must react. All wild animals, and indeed people too, have a well-developed instinct for survival. When an animal is threatened, this instinct takes two classic forms: fight or flight.

In almost all instances, a bison that feels threatened at the approach of a person will move away rather than fight, but if that animal perceives itself to be trapped, it may see no choice but to fight. Be sure to understand what a bison perceives as being trapped. What may seem an open area to you may indeed seem like a confined area to a cow with her calf. It could be that the animal just behind her, half an hour ago, attacked her calf in a dominance display. You did not see it, but she remembers and will not move in that direction. As a result she thinks her escape route is cut off, and to protect her calf she may choose to attack you. Like most

More people are injured by bison each year than by bears. Though people have learned to leave bears alone, they continue to approach too close to bison. Visitors to bison sanctuaries during July and August must be aware that this is the peak of the rut. During this time, bison can become very aggressive and may lose their normal respect for humans.

When visiting these special places, do not approach bison closer than 75 to 100 metres or 250–330 feet. To disrespect their personal space not only places your life in danger, but often results in the death of the provoked bison. In situations like this parking lot in Yellowstone, the bison has the right-of-way!

wild animals, bison cows are very protective of their offspring. In defending their calves, female bison have been known to attack other bison, horses, and people. Given the option, bison cows will always choose to walk away rather than risk a confrontation, but they will almost never abandon their calves, choosing instead to stand and fight.

By far the majority of bison attacks are caused by people, and most if not all of these could have been prevented had the people involved been aware of what to do when travelling in bison country. The ability to read and understand the signals given by an aggravated bison will greatly reduce the odds of your having a negative encounter.

Bison Bubbles

All animals, including people, have a personal space. This space varies in size depending upon your culture, your age, your sex, and the mood in which you find yourself. Some days you just simply do not want to be close to people. In a crowded store or standing in an elevator, you become very uncomfortable when others invade your personal space. In situations like the elevator you have no choice but to stand there and tolerate the intrusion. In more open environments you have the option to move away, so that the person who unknowingly violated your bubble is farther from you.

Bison too have a bubble, a zone around them that is theirs and theirs alone. They actually have three zones within their bubble that correspond to the level of pressure placed upon them by your presence, or the presence of another bison. These three zones are the *awareness zone*, the *escape zone*, and the *fight zone*. Where you are in relation to these zones determines the animal's response to your presence.

As a general rule, the size of the animals' bubble increases as they ascend the bison social ladder. Calves have the smallest personal space. The small space is essential to their survival, since they must stay very close to their mothers for the first few months of life. Calves are followed by yearling females, yearling males, subadult females, subadult males, and adult cows. These animals are the ones that form herds, and herd animals have smaller personal spaces than animals that live alone. Adult bulls, dominant bulls, and aged bulls, in contrast, live their lives mostly alone, or in small groups of two or three bison of similar age.

By recognizing which category of animal you are looking at, you can begin to understand its reactions to you, and as a result the animal becomes a little bit less unpredictable.

The Awareness Zone

Suppose you are travelling down a wooded trail interspersed with clearings. You come around a bend in the trail and there in front of you, several hundred metres away in a large clearing, is a lone bull bison, completely unaware of your presence. You decide to continue to toward him, and he eventually lifts his head, acknowledging your existence.

If you are a long way off, he will probably simply raise his head from grazing, look intently in your direction, then go back to eating. At this point, you are not a threat to him and he may continue his activities as though you were not there, even though you have now penetrated his awareness zone. The fact that he ignores you does not mean that he is unaware of you.

His reaction to your intrusion will vary depending upon the time of year and his recent life experience. During the breeding season in July and August, bulls tend to lose a bit of their natural respect for humans, and the size of their awareness zone is correspondingly smaller. During the winter, when a bull is experiencing significant energy demands just to stay alive, he may decide to move away before you even see him, just in case you are dangerous to him.

This cow was grazing quietly prior to the arrival of the photographer. She lifted her head and watched attentively, but was not alarmed to the point of moving away.

The Escape Zone

You continue a bit closer, passing through the bull's awareness zone and entering his escape zone. At this point, the bison has three choices: he can remain where he is in the hope that you are not going to come any closer, he can voluntarily move away from you, or he can try to move you away from himself. As with most wild animals, the option of leaving is the least stressful, so he walks away, around a small hill, and disappears from sight, thereby removing you from his escape zone.

By invading his escape zone, you have forced the animal to react. The level of reaction depends upon a variety of factors, and every encounter will be different and unpredictable. Given enough warning, the reaction will usually be a slow drifting off. If you arrive suddenly, however, his reaction could be much more spectacular: to explosively run away from you, or to turn and attack. Bison that bolt away rarely run far. They often run a short distance, not even sure what it was that startled them, but then turn and about-face in an attempt to determine what it was that came lumbering out of the forest at them.

Now is the time for wise choices on your part. This bison is alarmed, and should you continue to intrude closer, he may become dangerous. Choose a path that will keep you safe—a path that is at least 75 metres or 250 feet from the bull.

The Fight Zone

You continue your walk but decide to go around a small hill so that the bison can no longer see you. Unknown to you, the bull also decided to walk away, and suddenly you encounter each other at very close range. You now find yourself well past his escape zone and deep inside his fight zone. At this point he has no choice but to fight, because you are far too close and represent an immediate danger to him. The type of fight behaviour he exhibits at this point will be determined by a wide range of factors, such as the time of year and recent events in his life (did he just get beaten up by a larger bull?). His reaction could be as gentle as a few bluff steps in your direction, or it could be as serious as a full-blown attack.

In the unlikely event that you accidentally find yourself in a situation like this, as fast as you can get yourself to the nearest object of security: a large boulder, a tree, or your car. Once there, use it for protection. Bison are not relentless and will usually give up the fight as soon as you no longer pose a threat.

When travelling in remote backcountry areas, or even on busy front-country trails, always travel with your head up, watch where you are going, and remain aware of your surroundings. Being vigilant and seeing the animals around you ensures that you remain safe and have an enjoyable experience.

These two young bulls are getting ready for a fight. The bull on the left approached with his tail raised in a very determined and aggressive manner. The bull on the right then turned broadside to him, raised his tail even higher, and essentially told the newcomer: "Okay, I'm ready, bring it on!"

If you encounter the loser of a fight a few minutes afterward, he will still remember his loss, and if you crowd him, he may once again raise his tail, signalling to you that he is ready. If you see a tail position like that of the bull on the right and it is obviously aimed at you, then you have penetrated his fight zone. Depart quickly!

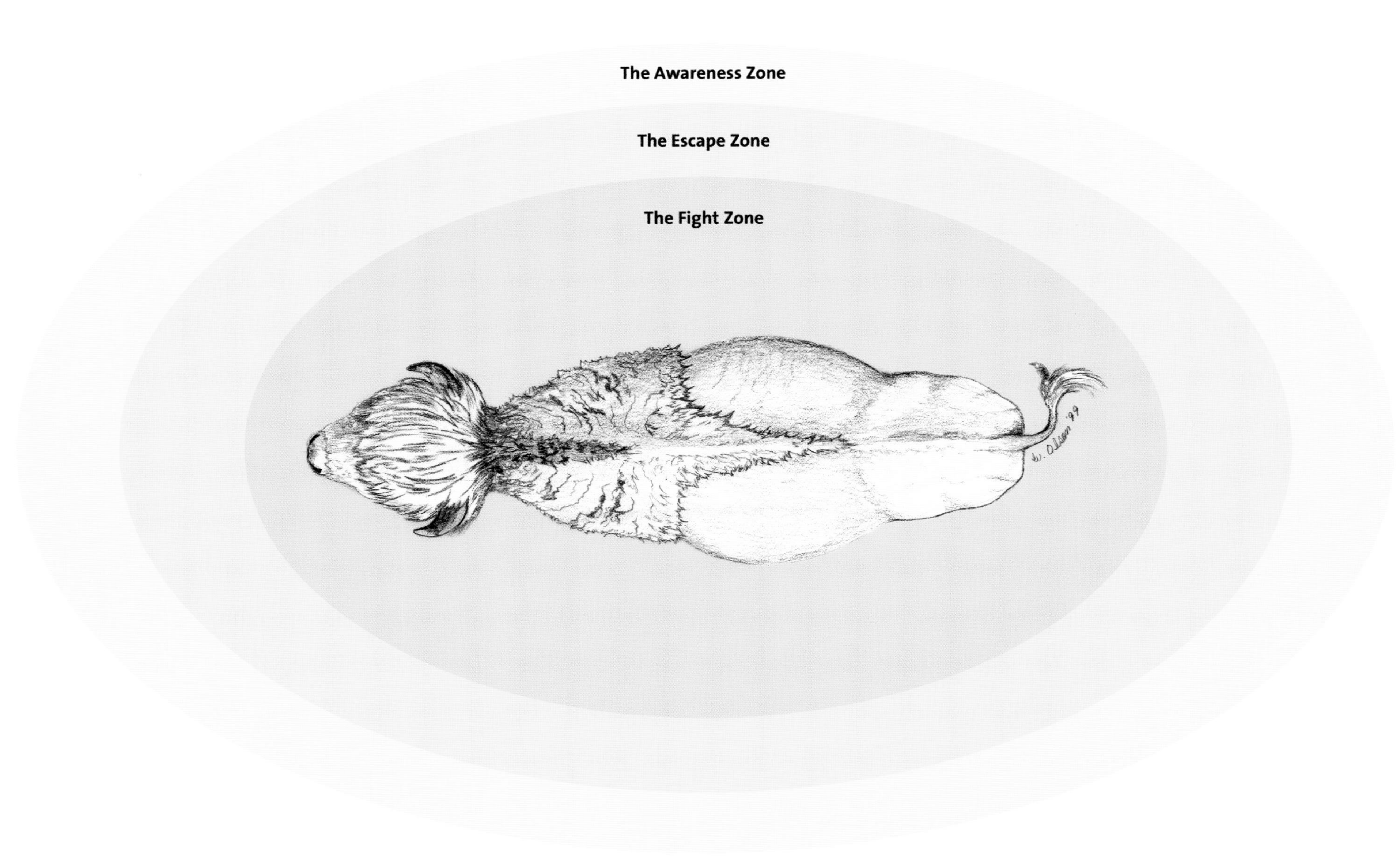

The bison bubble *is a concept all visitors to bison country should learn. There are no hard and fast rules about how wide these zones are, since they change from season to season, and indeed from day to day depending upon the recent experiences of the animal. It is most important to recognize that bison have a personal space, just as people do, and that you should not intrude within it. Try to never approach closer than 75 metres or 250 feet. This distance may put you inside the awareness zone, but should still allow you adequate time to leave if the bison becomes aggressive.*

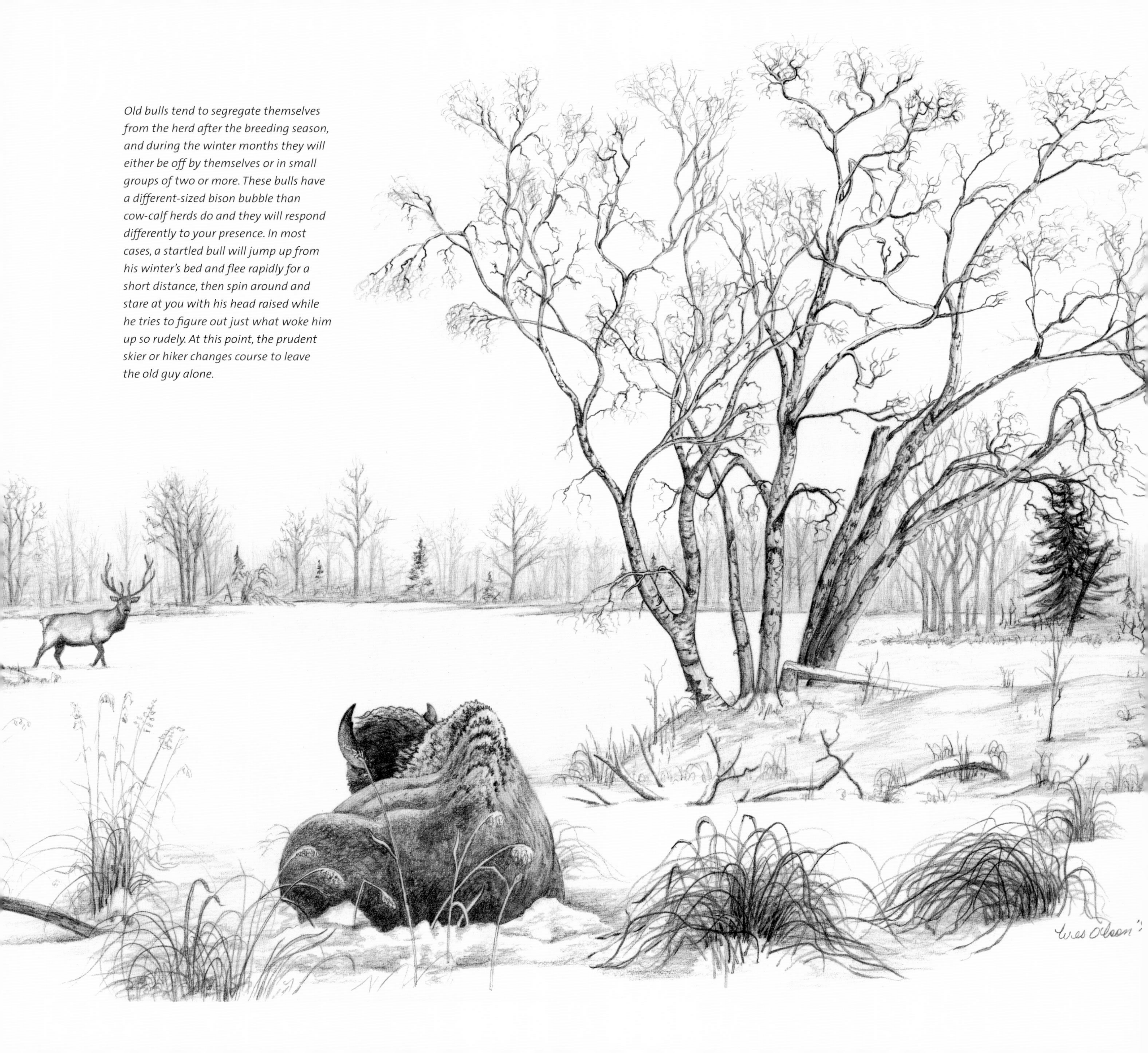

Old bulls tend to segregate themselves from the herd after the breeding season, and during the winter months they will either be off by themselves or in small groups of two or more. These bulls have a different-sized bison bubble than cow-calf herds do and they will respond differently to your presence. In most cases, a startled bull will jump up from his winter's bed and flee rapidly for a short distance, then spin around and stare at you with his head raised while he tries to figure out just what woke him up so rudely. At this point, the prudent skier or hiker changes course to leave the old guy alone.

Precocious and curious, calves are constantly being exposed to new and exciting experiences. They examine ground squirrels and butterflies alike with ears cocked forward and wet, black noses twitching in anticipation. For the first few weeks, they remain close to the protection of their mother's side, but by a month of age, the calves are venturing short distances afield in search of adventure.

Situations like this can be dangerous for the uneducated bison observer. Curious calves have been known to come close to investigate people, bringing with them an enraged and very protective mother.

What To Do Along the Roadside

IF YOU SEE a group of bison while travelling in your vehicle, follow these safety tips:

- If the bison are on the road and blocking your progress, do not honk your horn and charge at them with your car. Simply drive slowly toward them, and they will move out of your way. Bison attacks on cars are rare but when they occur, it is usually because someone became overly aggressive with his or her vehicle. This is particularly true during the breeding season in July and August.
- Stay in or close to your vehicle. Never approach bison on foot along the roadside.
- Bison will usually ignore your vehicle, but often spook if you stop and get out of it. If you must leave your vehicle, keep at least 75 to 100 metres or 250–330 feet away from the bison at all times.
- Avoid approaching bison in open meadows or clearings, where you have no escape cover.

Remember, you are accountable for your actions and behaviour around bison. By visiting bison sanctuaries, you are the intruder in their domain and you must govern yourself accordingly. Stay near your vehicle and do not harass the animals. This will ensure that the people who come along after you also have the opportunity to see and experience the bison.

Should You Cycle Past Bison?

IN THOSE PARKS AND RESERVES that permit the use of bicycles on roads and trails, cyclists should take extra care when encountering bison. A person on a bicycle travels quickly and quietly, and it is very important that you ensure the bison know you are coming. A sudden intrusion into their personal space makes a situation dangerous. If possible, move to the opposite side of the road from a herd, or better yet, wait for a car to escort you past the bison. On the trail, cycle around or carry your bike past the herd.

A herd of plains bison rutting along the parkway in Elk Island National Park has traffic stalled. Drive slowly through the herd and they will move out of your way. Should the herd run along in front of you, try to find a wide spot where you can get past. As soon as you manage to get by, the herd will usually stop running and return to their previous activities

What to Do in the Backcountry

Take extra precautions not to place yourself in danger when you travel on foot in the backcountry. Make bison aware of your presence by making noise. Bison startled at close range can be dangerous. As you approach a group or an individual bison, watch for the following signs as indicators that the bison are beginning to feel threatened by you. If you see any of these behavioural clues, retreat immediately.

- Is there a change in the bison's behaviour? Do they stop grazing and look at you; do bedded animals get up and begin to move away; or do any of the group turn and face you? If the animals exhibit any of these signs, then you have infringed upon their personal space.
- Watch for and recognize the signals given by the animal's tail (see page 30). The tail is a semaphore that accurately predicts the bison's mood, and it acts as your early warning signal that it is time to leave.

If you see any of the following behaviour, you are far too close and in imminent danger.

- **Head swinging back and forth while the bison stares at you**
  Head swinging is a normal behaviour displayed between bison, but it is not one that should normally be directed at a person.
- **Pawing at the ground or hooking it with the horns *in response to your approach***
  The important factor is the change in behaviour—does the bison change from grazing quietly to pawing or hooking the ground upon your arrival?
- **Short bluff charges toward you**
  An undisturbed bison never does anything quickly, so if you see a bison run toward you, assume things are not normal. These charges are usually very short and are employed by the bison to move you away. Bison constantly do this among themselves and are simply treating you the same as they would another bison. Just as another bison would, retreat promptly.

This young bull was grazing quietly until the approach of the photographer. He then stopped grazing, raised his head and stared at the person who penetrated his awareness zone. While the bison has not yet decided whether to flee or to fight, he is definitely aware of and focused on the intruder (top). *Inquisitive young calves may come over to investigate a photographer. Beware, a very protective mother might be right behind!* (bottom)

- **Loud, aggressive snorting**
  Bison are very vocal and constantly communicate with each other. They employ sound in their dominance interactions as a means of intimidating rivals, and if this sound is obviously directed at you, leave. The important distinction here is whether that sound is directed at you or at another member of the herd.
- **A determined advance straight in your direction**
  If a bison is heading straight for you, find adequate cover quickly. Never place yourself so far from cover that you will not have time to get to it. Walking across a large clearing that you know contains bison may be a rather costly mistake if you cannot get to the trees in time. Bison can easily outrun a horse and have been clocked at over 50 kilometres per hour 30 miles per hour.

  If you suddenly find yourself face to face with a mad bison, get out of its personal space as fast as you can, do not make eye contact, and look for any form of escape cover—a tree, large boulder, anything to put between you and the aggravated bison.

  Once you have safely put some distance between yourself and the bison, remember the clues you saw the bison display, and try not to place yourself in a similar situation in the future.

As a general rule these guidelines should enable you to travel safely through bison country, but they are by no means a guarantee of your safety. Most people injured by bison asked for it. It is not safe to pet a large bull on the forehead nor to get close to a newborn calf, so do not try. Bison can be unpredictable and every situation will be different. Use common sense, and above all, do not penetrate their bubble!

A lone bison bull grazes along the edge of an aspen forest in Elk Island National Park. Bulls like this require lots of space, so give them a wide berth.

Set Your Sights on Safety

Bison tails are seldom still. Constantly on the move to keep flies and mosquitoes off the sensitive skin of the hips and flanks, the tail of a bison is also a semaphore of its intentions. Four basic positions transmit the body language of the bison.

Position 1. When the bison is standing at rest, the tail is held in a comfortable position, usually slightly away from the body.

Position 2. This position indicates interest or curiosity. Something—another bison or perhaps you—has captured the animal's attention, and usually it will be looking directly the object in question. You may have just entered this animal's awareness zone.

Position 3. When the tail is held above the horizontal, you know that the bison's level of tension has increased. This tail position is commonly used to displace another bison from a wallow or when a breeding age bull approaches a cow in heat.

Position 4. You do not want to be the cause for this signal. If you see the tail raised past the vertical, something is about to happen and it is usually dramatic and explosive. The tip of the tail rapidly flags back and forth, the back is humped and rounded, and the head is usually placed close to the ground. This is either the signal between two rival bulls that battle is about to commence, or one given by a cow defending herself or her calf.

This tail position is similar to that displayed just prior to a defecation, but in that instance, the other aggressive signals are absent.

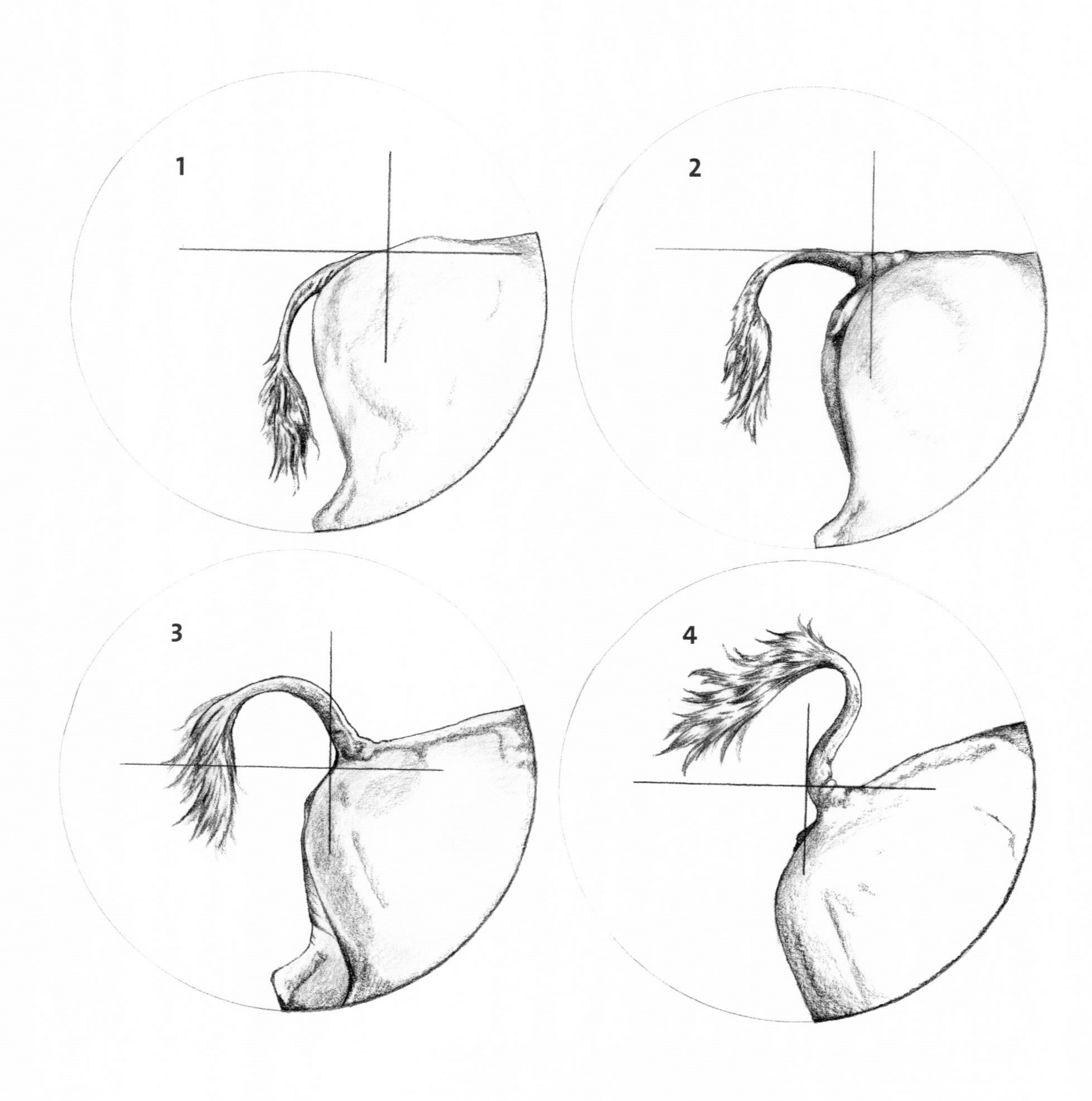

"There's only two reasons why a buffalo raises his tail. The first is to charge, and the second is to discharge," said Vern Ekstrom of Custer State Park, South Dakota. In neither case do you want to be on the receiving end.

This brown-headed cowbird is catching a ride, and perhaps supper, on the back of an old bull. The cowbird evolved to follow the migrations of bison across the landscape, foraging on the insects disturbed by their passing hooves. Because it is constantly on the move, it cannot take the time to stop and raise a brood. The brown-headed cowbird is known as a parasitic nester, since it deposits its eggs in the nests of other bird species. The eggs are then incubated and the young raised by the unsuspecting mother, often to the detriment of the original young.

3

Seasonal Bison Social Structure

Wolves like this one, almost obscured by shrubbery, are efficient predators on bison where their ranges overlap.

A BISON POPULATION'S social structure is complex and constantly changing. It is difficult to keep track of the comings and goings of various individuals or subgroups, but with enough time and patience, you can clearly see the shifts in population structure over the course of the seasons.

Like flocks of swallows, bison herds are constantly splitting and rejoining, picking up new members from one group, and yielding some to another. If you were able to remain with a bison population throughout the course of a year, definite patterns of social structure would be observed. Many factors influence the dynamics of bison social structures, and depending upon the location of the herd there may be variations from the patterns described in this chapter. As a general rule, however, these patterns provide a guide to help you interpret and understand bison social dynamics.

Winter

TWO PRINCIPLE TYPES of bison groups exist during the winter months, up to about the end of March. Most common are the mixed herds of cows, calves, yearlings, two-year-olds, and three-year-olds. Their main objective is to survive the harsh realities of winter. Extreme temperatures, deep snow, and, in some regions, the ever-present risk of encountering a hungry wolf pack force bison to form groups large enough to provide the required protection. These groups are called *matriarchal bands*, since they are often led by one or more older cows. Most of the adult cows in these bands will have been bred during the previous summer, and their pregnant state requires the cows to select habitats that will provide them with the nutrition needed to carry their calves to full term. The combination of large group size and high nutritional requirements results in the band seeking out large meadows of relatively high-quality forage.

The second group of bison is composed entirely of bulls. These bull groups are known as *bachelor bands*. After the breeding season is over in the late summer, the mature bulls leave the other members of the herd and disperse across the winter range, often forming small groups of similarily aged bulls. Frequently bulls will remain alone and separate from the herd throughout the year, and they are known as *loner bulls*. Three-year-old bulls leave the protection of the matriarchal band at this time since they now have the pelage characteristics of a mature bull, and should they remain with the matriarchal band, they could be selected by predators because they look different from the others.

Matriarchal bands—mixed groups of cows and juveniles—seek the shelter of forests, wooded river banks, and ravines in an attempt to mitigate the blast of winter's fury. Comprised of several generations of bison, these small bands disperse across the landscape, seeking wetland meadows and patches of habitat large enough to support the group over several days of feeding.

During periods of severe cold, such as this −30°C or −22°F day in Elk Island National Park, bison continue to graze on frozen sedges and grasses adjacent to the region's wetlands. Other ungulates such as elk, moose, and deer bed down and remain motionless as a means of conserving vital body heat. This photo shows a bachelor band of similarly aged adult bulls (inset right).

Spring is also a difficult time for the yearlings. Weaned both from milk and the security of their mother's sides, these young animals are cast adrift in bison society. They temporarily form yearling bands: groups of bison of a similar age. The bonds formed during this period may last a lifetime—which for bison may be a quarter of a century or longer.

Bison undergo two hair moults each year: one in the fall when they grow their winter coat and another in the spring when they shed that coat to grow a lighter coat for the summer.

Spring is a difficult time for pregnant cows. Over the winter, they will have endured months of extreme temperatures, deep snows, and the nutritional stresses imposed on them by their growing fetuses. In the spring, the cows seek out habitats with the richest, most abundant forage in an attempt to satisfy their energy requirements. It is at this time that the pregnant cows may temporarily abandon their yearlings in preparation for giving birth to their new calves.

Spring

ONE OF THE MOST traumatic times in the life of a bison is spring weaning—not weaning from mother's milk so much as weaning from maternal care. When the pregnant cow comes near to calving, she forcibly abandons her yearling so that all of her attention can be placed upon her newborn. This happens in all matriarchal bands, and within a short while small groups of yearlings, known as *yearling bands*, form. Often older, barren cows associate with these yearlings. While the existence of yearling bands is relatively short (the duration of the calving season, approximately two months), the bonds that develop between yearlings during this period are very strong and can last a lifetime. The bond is most noticeable between females, but also exists to a lesser extent between males.

When a cow is ready to calve she may either remove herself from the herd and seek seclusion to give birth, or she may simply move to the edge of the group. Often the first cow to calve is an older cow, one who entered the previous breeding season in better condition than her cohorts. This is a dangerous time for her and her calf in regions populated with wolves, since predators will immediately see and key in on this newborn. The cow then has a choice—she can remain with the herd and hope for communal protection, or she can seek solitude and hope to be overlooked. Even in Elk Island National Park, where coyotes are the only predators, this behaviour can be seen, so strong is the bison's instinct to protect their calf.

Once calving is underway these new mothers quickly form groups known as *nursery bands*. A caste society develops of the "haves" and the "have-nots" with the cows who have calved segregating themselves from those who have not calved. The new mothers are very protective of their offspring, and the youngsters cling to their mother's side for the first week of life.

As the end of the calving season approaches, another group of bison can be observed: the old or barren cows that will not produce a calf that spring. These groups are known as *spinster bands*. Their existence as a discrete social group is relatively short and easily overlooked. As the calves age during their first few weeks of life and their mothers begin to relax a bit, the nursery bands begin to accept other members of bison society into their group. Within a short period of time, the old cows are once again mixed into the herds and the matriarchal bands reform, just in time for the breeding season.

This small mixed group of plains bison in Elk Island National Park is typical of matriarchal bands. Once calving is complete, the cows permit the return of the yearlings, and the group once again is composed of all ages and sexes. The noticeable exception are bulls older than about three. They will not join the matriarchal band until the beginning of the rut, in mid-July (above).

Bison evolved to produce their calves in a synchronized, short calving season. The sudden production of large numbers of identical-looking calves confuses predators and makes it difficult for them to select one individual from the herd. Once calving is complete, groups of the rambunctious youngsters can often be seen playing together. These nursery bands are often accompanied by one or more cows, which may or may not be the mother of one of the calves (right).

Calves look alike at birth. They are all the same size, and reddish orange in colour and very precocious (below).

Bulls wander through the herd, stopping briefly to check the scent of a cow's urine to determine whether she is in heat and receptive for breeding.

They do this through a behaviour common to large mammals, known as "flemen" or the lip curl. A male takes a good whiff of the female's urine, then tilts his head back with the upper lip lifted. This exposes the molecules in her urine to the receptors in his nasal passages and lets him know whether she might be close to breeding. If she is not ready, he moves on to the next female, and the next, until he finds a receptive one. This constant searching and tending requires dedication by the bull—so much so that he may not eat for days at a time.

Summer

Whenever two or more matriarchal bands come together, they form a *cow-calf herd*, and this is the grouping most commonly observed by visitors to bison preserves throughout North America.

The next most common grouping is the *breeding congregation*, which occurs during summer when cow-calf herds and bachelor bands unite to form the largest herds. Back when the vast herds of millions roamed North America, the tumult and noise associated with the summer breeding congregations must have been overwhelming. The constant roaring of thousands of males, the clicking and clacking of hooves, and the soft grunting of calves and cows amid a choking pall of dust and flies would have left the human spectator feeling a tad small and insignificant.

Those huge herds were a temporary phenomenon, however. They were formed by countless other, much smaller groups coalescing and melding into vast herds numbering in the tens of thousands. These breeding herds were constantly on the move, since it takes a lot of forage to support this many animals. They could not stay in one place for long before they had to move on in search of fresh pastures and water. This constant moving of herds from one area to another resulted in different herds mixing together, then splitting apart again, with bulls from one herd leaving to fight a rival in another and young bulls leaving the matriarchal bands for the first time.

The peak of the breeding season lasts for about a month, during the last two weeks of July and the first two weeks of August, with a two-week shoulder season on either side. Once breeding season is over, these huge congregations split up again into smaller groups, enabling the animals to stay in one place for a longer period of time.

Summer is a time of courtship, a time when breeding-age males dedicate themselves to seeking out and tending cows that may be receptive to their advances. During the rut, bulls circulate throughout the herd, examining each cow in turn in an attempt to learn whether she may be approaching her estrous (period of heat). If she is, he will often spend several days by her side, departing only briefly to defend her against the attentions of rival males.

Down from the hills for a morning drink, this herd in Yellowstone's Hayden Valley is typical of the cow-calf herds that form prior to the rut (inset above).

Wallows

When visiting bison territory, have you ever noticed large circular patches of exposed soil and wondered what they were? Once the prairies of North America were dotted by thousands of them, all created and maintained by bison. These are wallows, the equivalent of our coffee shops—a place to socialize, to gather, somewhere to go to get to know the neighbours.

To bison, though, they are much more than that. Wallows are used by all members of bison society for a variety of purposes. Perhaps the foremost of these is the dust bath, which is essential to the health and well-being of bison. When a bison thrashes and rolls in the dust of the wallow, the dust is ground into its hair, down deep against the skin. This layer of earthy material provides protection from the harassment of biting insects such as mosquitoes. It also helps to protect bison from cold during rainstorms. When the packed dirt in the hair gets wet, it helps to shed the rain and thus provides an added measure of warmth.

Rutting bulls use wallows extensively: to intimidate rival bulls, as a dominance tool between other members of the bison group, and as a place in which they apply that most alluring of bison perfumes—urine! Bulls will often hook at, paw at, and thrash in the dust of wallow, then leap to their feet and urinate in the resulting dust. Once these essential preparations are made, the bull then rolls in the irresistible concoction and once anointed, goes courting. The urine both carries with it the scent of his testosterone and is a clear message to females that he is the mate of choice. It also signals rival bulls that he is bigger, meaner, and not to be messed with.

As a visitor to bison country, take the time to visit a vacant wallow and try to see who was there ahead of you. The fine dust left behind will yield fascinating clues such as tracks of bison, coyotes, or other wild animals, along with hair and animal droppings.

Thrashing and rolling in an exuberant display, this bull is achieving several goals at the same time. He is applying a liberal dose of mosquito repellent, a raincoat, and perfume, and perhaps also scratching where it itches.

Fall

IF YOU VISIT A PARK in the fall, after the breeding season is over and the leaves have begun to turn, you will notice that the size of the average bison group is fairly small. It varies from park to park, depending upon the environment and habitat, but as a general rule, at this time of year the larger summer groups splinter into small herds of fifteen to twenty animals including adult cows, their calves, and female yearlings. Mixed in will be some young bulls, but few older than three years. It is unlikely that any adult bulls will be in the group unless there is a cow in heat. The matriarchal bands are led by a lead cow, usually one who is fairly old, but not necessarily the oldest in the group. She may well be the mother of several members of her group, or the band could be composed entirely of strangers, come together by chance. In some parks, like Elk Island National Park, these groups will often have two or three cows of the same age: cows that were abandoned by their mothers as yearlings and that have formed a life-long bond that makes them nearly inseparable.

With the end of the rut, the bulls leave the summer range and seek a place to winter. They may travel several hundred kilometres or miles, exploring and occupying new habitats. Historically, there was a seasonal shift on the Great Plains from the huge summer aggregations, to smaller groups occupying the fringe areas. Often this was a shift toward wooded areas, where they sought refuge from winter storms and found forage not consumed during the summer months.

Fall is a critical period for the breeding bulls. During the rut they spend most of their time tending and breeding cows, with a resulting weight loss, since they essentially stop eating for an extended period of time. If they are to survive the coming winter, it is essential that they regain as much of their lost weight as possible and build up their hair coat and winter fat reserves.

Bison that are dominant breeding bulls tend to forget to eat during the rut. They have other more important things on their minds, and as a result they may lose a tremendous amount of weight. It is vital to their survival that they gain back as much of their lost weight, prior to the onset of winter, as possible.

This thin, old bull has found a patch of short grass that is higher in nutrients and energy than the taller, coarser grasses behind him. He will continue to search out these grazing lawns throughout the fall and early winter, but once winter's snow arrives, he will switch to sedges and taller grasses (bottom right).

A bison cow and her calf rest quietly after an afternoon of feeding. Like cattle, bison have four stomachs. The forage is first plucked from the ground by incisor teeth designed to graze close to the ground. The food is quickly masticated and saliva added to assist the initial stages of digestion. It is then swallowed and enters the first stomach, where the food remains until the bison has the opportunity to rest quietly. Then, the bison regurgitates the partially digested food. This mass or bolus of food is referred to as the cud, and once the bison has finished grinding it into a fine mush mixed with lots of saliva, it re-enters the digestive system.

The food a bison consumes stays within its digestive system 30 percent longer than in cattle, and thus the bison is able to subsist on poorer quality forages, or if on high quality forage, to consume 30 percent less of it.

Wes Olson '96

It is mid-afternoon and the herd has come to rest on a hillside in the National Bison Range, Montana. Bison have evolved to produce their calves in a very short period in an attempt to flood the prairie with new calves. This synchronized calving strategy is designed to overwhelm predators and offer newborns the best chance of survival. The new calves' dun-coloured coats blend in with the dry spring grasses, and when stationary, they can be very difficult to observe.

Unlike deer, which hide their young, bison calves are "followers." From only a few minutes of birth, they are able to travel and keep up with the herd. During precontact times, herds of hundreds of thousands roamed the plains. These large herds could not stay long in any one place because they would quickly consume the forage around them, forcing them to continually move from place to place in search of food and water. Calves that lagged behind quickly became food for hungry wolves.

4

The Who's Who of Bison Society

Hayden Valley, Yellowstone National Park.

Bison society, like any other society, is made up of individuals, each with his or her own personality and characteristics. Bison societies are hierarchical in nature with every animal occupying a specific place in the social system. Their place within the society is never static and may change frequently, depending on the other bison encountered. As the individual ages and matures, it tries first to ascend the social ladder and second to maintain its position there. This is particularly true of males during the breeding season, but it applies to all bison during the course of their lives.

Without a clear understanding of who these bison are (as determined by their physical appearance), it is impossible for researchers to learn more about the social dynamics of a bison population. Once the bison enthusiast has the ability to look at a bison and determine its age and sex, he or she can then begin to learn more about how that individual fits within the society, how it interacts with other members of its group, and how those groups interact with other groups. Your understanding of a bison population's structure will yield valuable clues as to the health and vigour of the herd. Reproductive rates, recruitment rates, and mortality rates are all indicators of how successful a bison herd is at surviving the stresses placed upon it by everyday life in the wild. Your ability to assess each of these factors is enhanced by your ability to recognize the age and sex of the members of the bison community.

The illustrations and photographs here depict the characteristics of bison during the summer months. There are considerable pelage changes with the arrival of winter, and all ages and sexes look different with their winter coats on.

This small herd of plains bison typifies the subspecies as they existed for thousands of years. Plains bison developed physical traits that exaggerated their appearance in profile. These well-developed physical and pelage characteristics permitted a breeding-age bull to stand out in the crowd—and in crowds numbering in the hundreds of thousands, this was essential if the bull wanted to pass on his genes.

Dozing quietly on a warm hillside meadow, this calf requires frequent naps to replace the energy lost through its wild cavorting and gambolling with other calves, butterflies, ground squirrels, and other new and wondrous creatures (right).

Calves
(from birth to 12 months)

Social Status and Behaviour

After a gestation period lasting about nine months, the majority of bison calves are born during May and June. The calving period differs slightly from one area to another and may begin as early as April and end as late as September or October. The cow usually leaves the group a short time before she is due to calve, with the actual birth taking about twenty minutes to complete. Newborn plains bison calves weight from 14 to 23 kilograms or 30–50 pounds. After about half an hour, they are able to stand, and within one to three hours, they are able to run and keep up with the herd. Unlike the deer family, which hides newborns as a means of protecting them from predators, bison calves have evolved to flee from danger, as a means of escaping it.

By the time a calf is four to six hours old, it begins to take sample nibbles of the grasses that will sustain it later in life, but mother's milk will supply most of the nourishment it receives for the first four to six months. During the first two or three weeks, calves are totally dependent on their mothers for food, protection, and education. After this, however, they become a bit more adventurous, and a calf may move a considerable distance away from the security of its mother's side. Calves often form nursery groups numbering from two or three to a dozen or more. These bands of rollicking youngsters are usually tended by one or more cows that might not have a calf of their own. It is common to see two calves together, but actual twins are rare. About the only way to determine whether a pair of calves are actually twins is to witness them nursing from the same mother.

During the first few weeks of life, there is little or no dominance hierarchy among the calves, but by the time they reach about four months of age, male calves become dominant over female calves. Within each sex, the most aggressive animal rather than the largest becomes dominant, and an aggressive female calf can often dominate a larger male calf. Social dominance may be taught to the calf by its mother through the simple process of association. When a dominant cow approaches another bison that is lying in a wallow and displaces it, this action is observed by the dominant cow's calf. Because the calf is constantly at the cow's side, it has the opportunity to witness these actions repeatedly. Even at an early age, whenever it can a calf may begin to assert its dominance over its peers.

Colour Phases of Calves

The umbilical cord dries to a stiff cord within the first three days of life and drops off after about one week (top left).

Snug in the shelter of her mother's chin, this calf will not stray far from her mom's side for the first week to ten days (bottom left).

After they develop a bit of confidence, bison calves routinely romp and cavort together or alone. They are a follower species and almost from birth, they are up and able to follow the herd (bottom right).

From birth until about eight weeks of age, bison calves are reddish orange—a colour that lets them blend into the sand-coloured grasses of early spring so they can better escape predators (top left).

By about ten weeks of age, the calves begin to turn a mottled chocolate brown colour as their baby coat is gradually replaced with the darker colour more typical of yearlings (top right).

By the time calves reach twelve to thirteen weeks of age, around mid-August to early September, they have changed colour totally and lost the last remnants of their baby coat. The new chocolate brown colour allows them to blend in with the rest of the herd and thus makes it more difficult for predators to identify them as potential prey (bottom right).

Male Calves

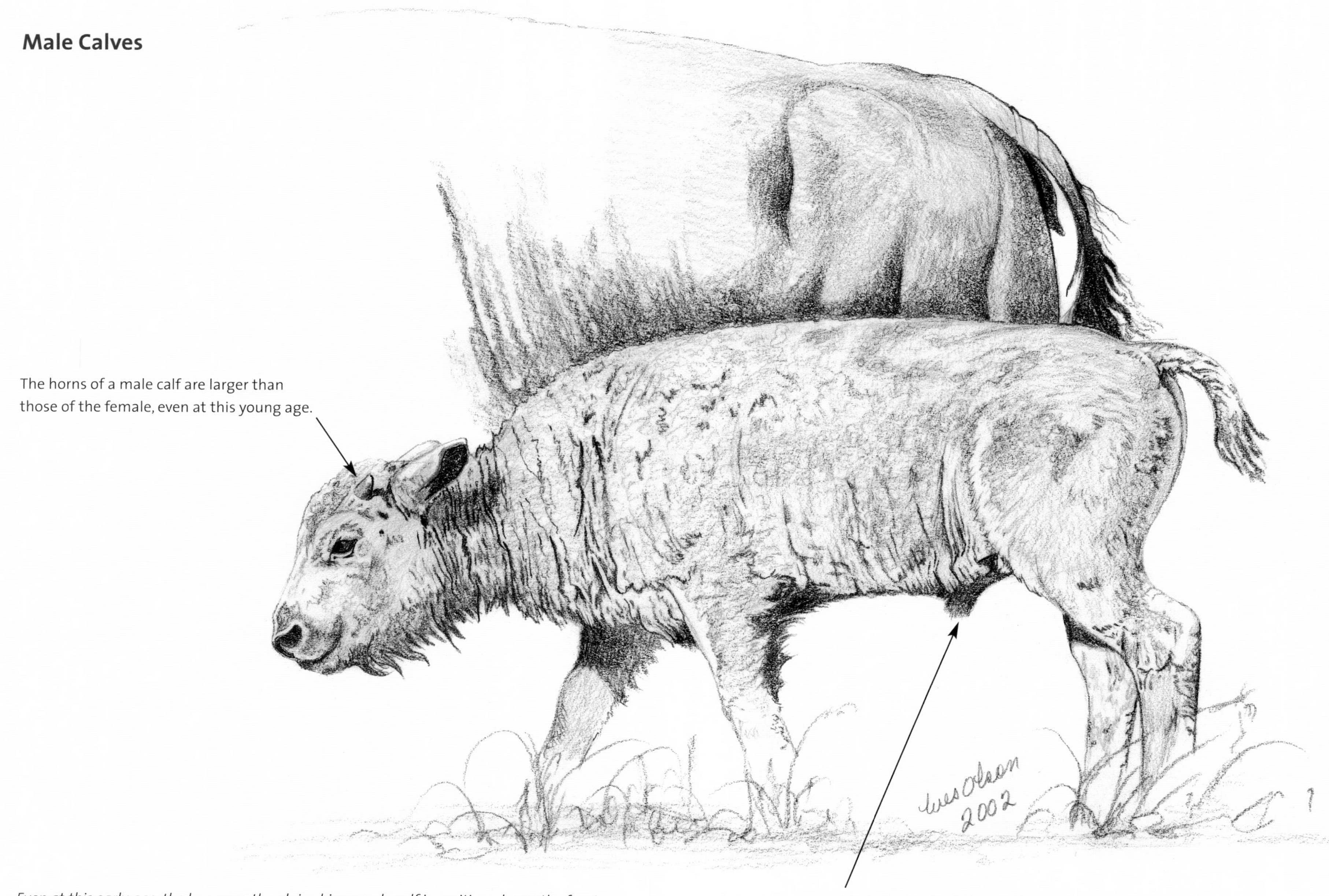

Even at this early age, the hump on the plains bison male calf is positioned over the front legs. His horns begin to develop within a few days of birth, and by the time he is three to four months of age it is possible to determine his sex, based upon the size of his horns. Calves are born in a roughly 50:50 male to female ratio, but because adult males tend to expose themselves to greater risks, such as fighting, exploring new territories, and during the winter, occupying poorer quality habitats, the adult male to female ratio is often much lower. Calves are sexually dimorphic at birth, which simply means that males weigh more than females.

Female Calves

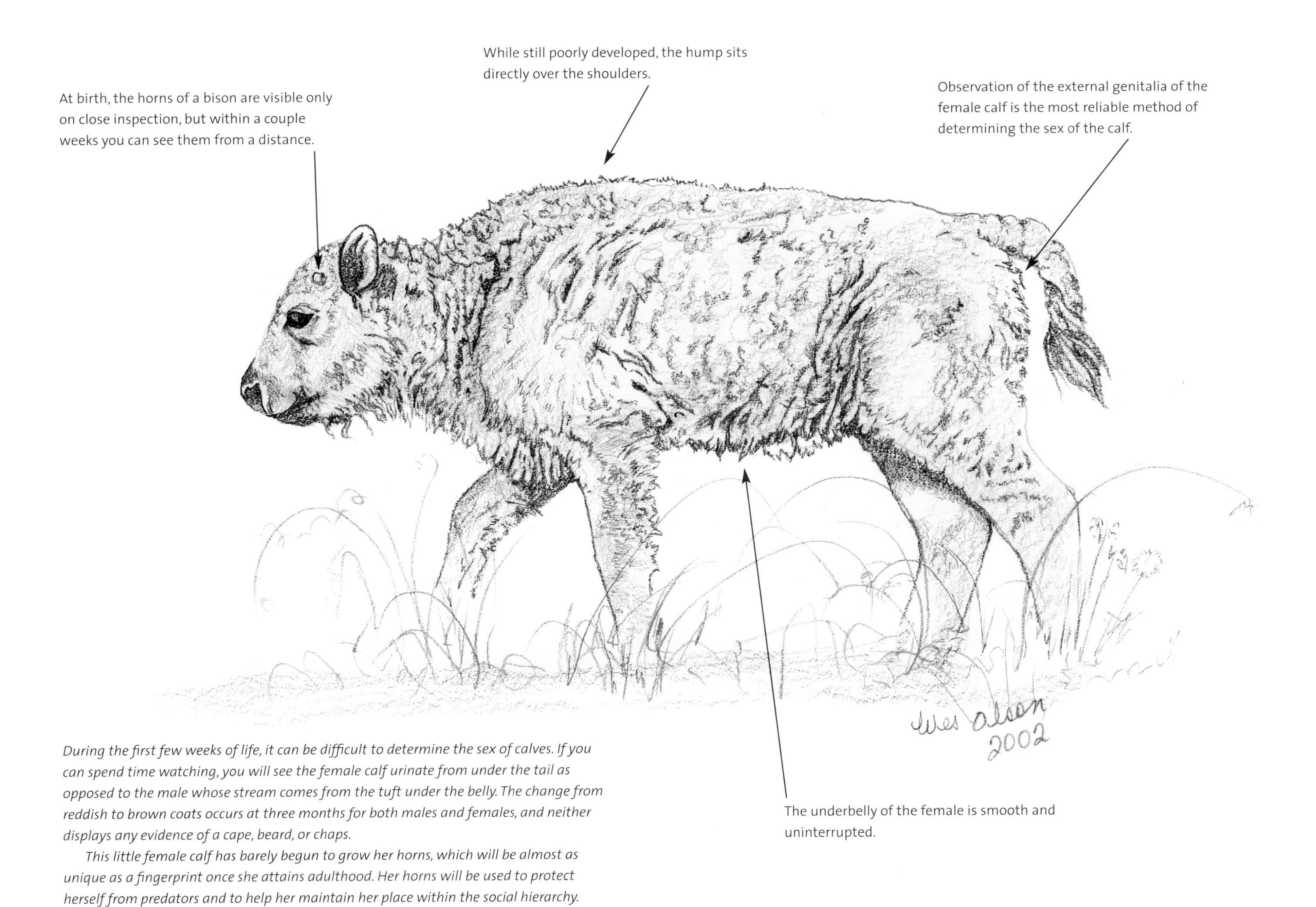

During the first few weeks of life, it can be difficult to determine the sex of calves. If you can spend time watching, you will see the female calf urinate from under the tail as opposed to the male whose stream comes from the tuft under the belly. The change from reddish to brown coats occurs at three months for both males and females, and neither displays any evidence of a cape, beard, or chaps.

This little female calf has barely begun to grow her horns, which will be almost as unique as a fingerprint once she attains adulthood. Her horns will be used to protect herself from predators and to help her maintain her place within the social hierarchy.

Yearlings (from 12 to 24 months)

Social Status and Behaviour

Yearlings of both sexes fall next to calves on the bottom rungs of the social ladder. They can often be seen exercising their limited dominance over the calves, but they do so at their own peril because cows can be quite protective of their young. During the period just before calving, usually during the last part of April or early May, you can often see fairly large groups of bewildered yearlings. Prior to calving, the cow severs the bond with her yearling. Deprived of their mother's protection, the yearlings appear confused and lost. All of these now "homeless" yearlings band together out of, perhaps, a shared sense of abandonment. The yearling band often has one or more older cows mixed in with them, and to the untrained eye, it looks a bit like a group of cows with a few bulls thrown in. A closer examination reveals that the sizes are wrong and that something else is going on.

This is a very insecure period in the life of a bison and a time fraught with danger. The yearlings have lost the protective shield behind which they have sheltered, and they must now begin to fend for themselves and establish their place within the society. I often wonder whether the presence of barren old matrons is because of the young animals' desire for adult companionship or because of the cows' desire to be close to young bison. It is more likely that the matrons are plagued by the youngsters, and I can easily picture these spinsters attempting to leave the youngsters behind, only to look back and find them following at every turn. Eventually these matriarchs must resign themselves to the presence of the bothersome things and begin tutoring them to act in ways more suited to bison.

The yearling female (right) *has ears roughly equal to or slightly shorter than her horns, which are shorter and of smaller diameter than those of her male counterpart* (left).

Male yearlings can be particularly obnoxious during this period, comparable to teenage males in human society. They can frequently be observed trying to establish dominance over their younger siblings and even over older animals. Physically, they resemble two-year-old cows although they are considerably lighter in weight. Pushing and shoving matches among male yearlings are quite common but present no real physical danger, in contrast to the older age groups.

On the social ladder, yearling females occupy one rung higher than calves and usually one rung lower than yearling males. Animals in this group that are in excellent physical condition with above-average body weight may participate in the breeding season. The percentage of yearling females that become pregnant, however, is very low.

Both sexes continue to nurse for as long as their mothers allow them to do so. If a cow does not conceive during the breeding season, she may be content to let her yearling nurse well into its second summer. This is unusual, however, as most calves are weaned by the time they are six to eight months old. Under severe conditions—extreme summer drought, for example—the cow may not be able to produce milk even in the summer and the calf may not survive until fall.

The ears of both sexes are shorter than the horns, and in both they are easily discerned from the hair on the head. The plains bison yearling bull may have begun to grow the chaps on its lower legs, but they will not be very obvious.

Male Yearlings

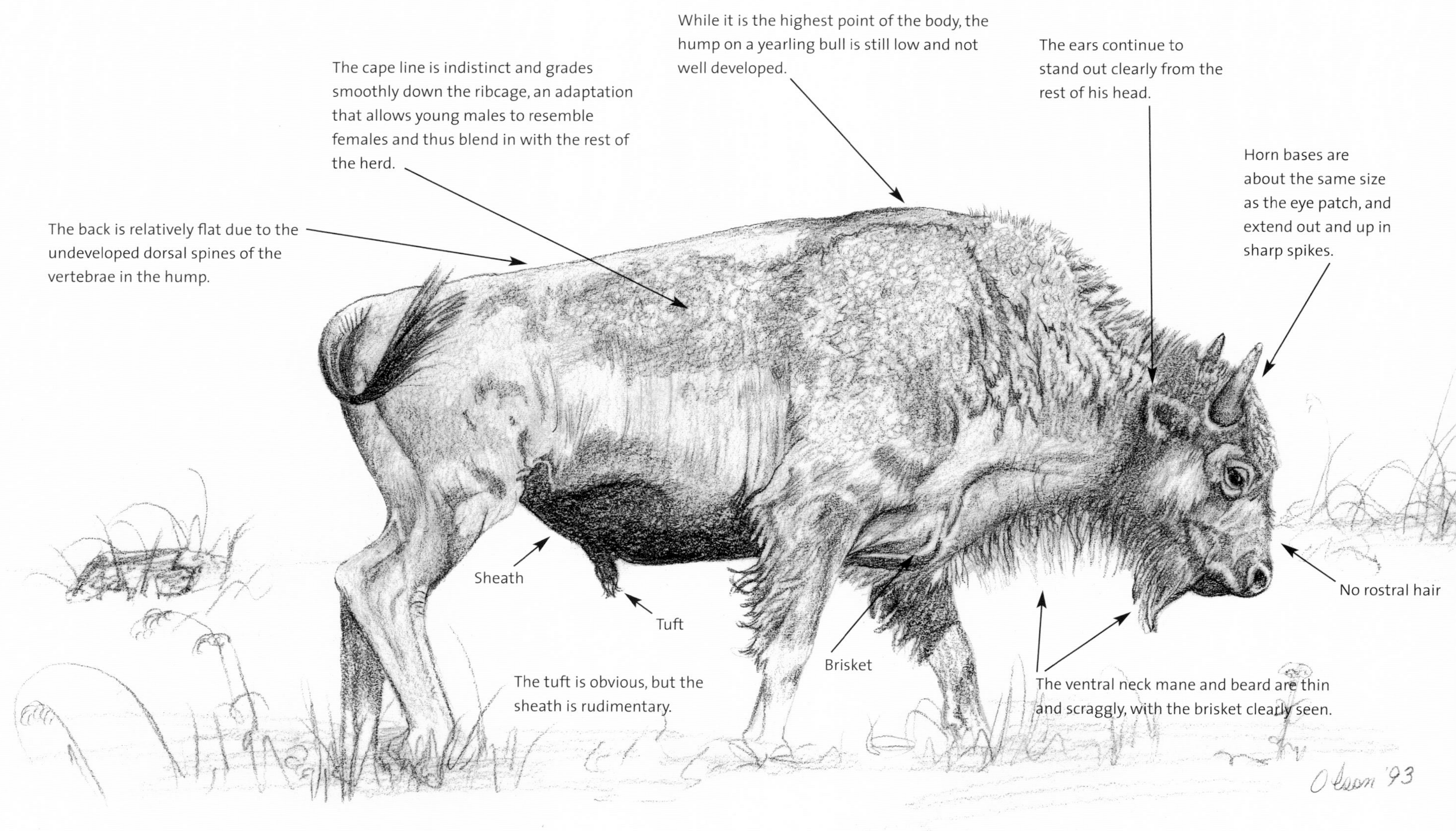

The yearling plains bison bull resembles subadult cows both physically and mentally. He occupies a rung on the social ladder that is fairly low, and he is easily dominated by older males. While still slim, his horns are thicker at the base than are those of the yearling female. By this age, the horns project out and up past the ears, with little or no recurve at the tips. His cape grades smoothly over the ribs, and his beard is thin and scraggly. The highest point of the hump is over the front legs.

The top of the yearling plains bison bull's hump is roughly equal in height to the base of an adult cow's tail. His back is relatively flat due to poorly developed dorsal vertebrae, which will later form his hump. His chaps are rudimentary and tend to look like those of a young cow. He can now be easily identified as a male by the presence of a tuft, but his sheath remains poorly developed. His chaps have finally begun to grow, but they are shorter than the leg is wide. Not until he reaches adulthood at the age of seven or eight will this bull begin to emit the spectacular vocalizations so characteristic of rutting bull bison.

Female Yearlings

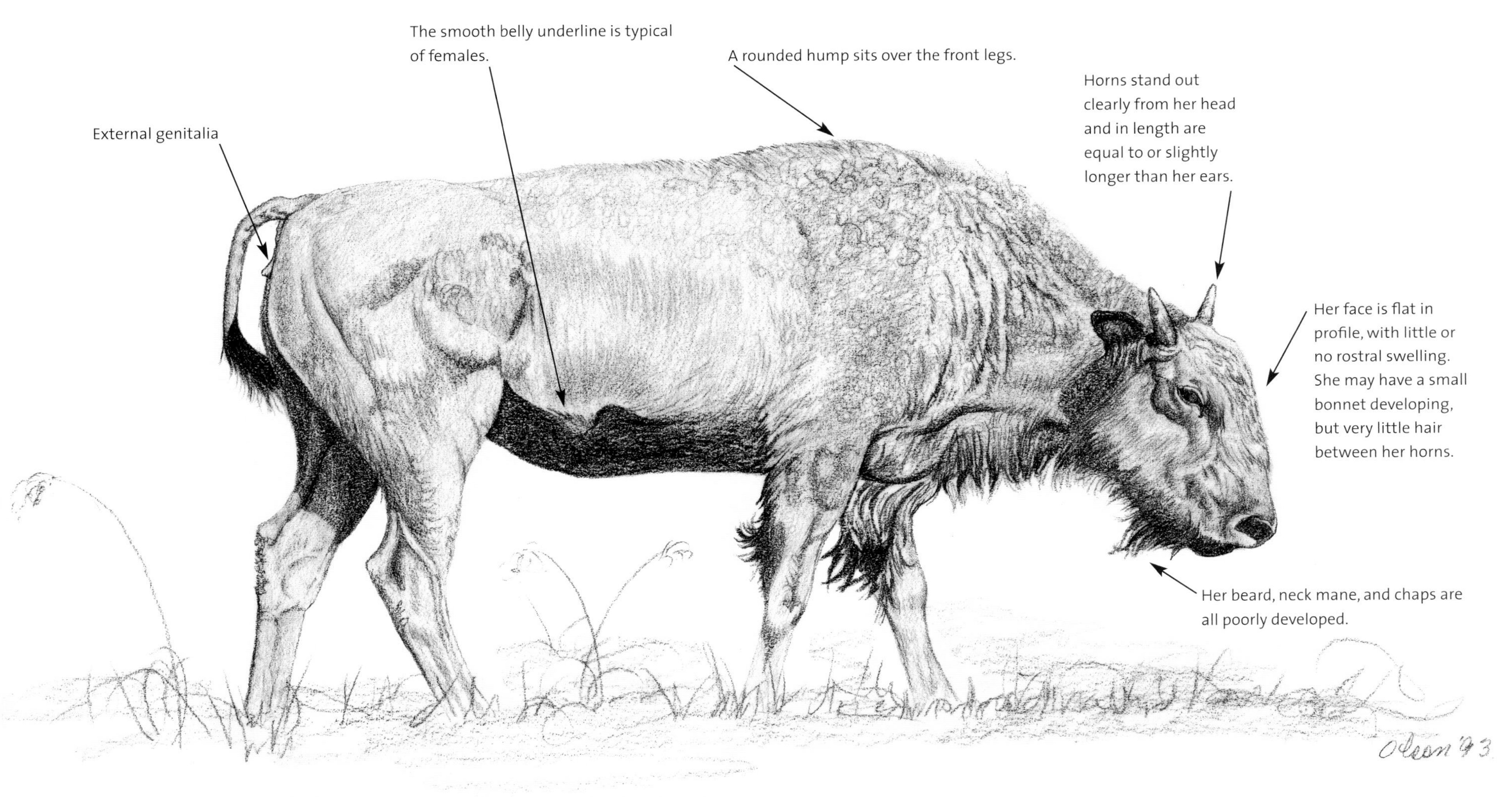

As with the male yearling, the female plains bison yearling's hump is also highest over the front legs. Her ears stand out very clearly from the rest of her head, and her horns are fine and delicate. The hair between her horns is very short, and there is little evidence of the bonnet that she will wear as an adult. Occasionally a yearling female will be large enough to conceive during the breeding season, but this is rare in the wild.

The yearling female plains bison has a sleek and uninterrupted underbelly. She shares many physical traits with the yearling male, but the finer horns and the different external genitalia easily distinguish the two.

With his hair all matted and wet from a recent rain, this young wood bison bull looks a little bedraggled and worn. An important function of buffalo wallows is to provide a raincoat for bulls such as this one. When they wallow and thrash in the dust of a wallow, the dirt and dust gets packed into the hair-coat, and when wet it forms a covering that stops water from penetrating to the skin. This keeps the bison warm and dry even in the coldest of rainstorms.

Two-Year-Olds (24 to 36 months)

Social Status and Behaviour

Two-year-olds occupy the next rung on the bison social ladder, just above the yearlings. The two-year-old bulls mimic the pelage characteristics of subadult cows, with the cape extending back along the flanks toward the hips. While still poorly developed, his chaps and neck mane have begun to grow; this feature can be used as an indicator to separate male two-year-olds from male yearlings.

This is the last year that the two-year-old bulls will stay with the cow-calf herds and take advantage of the benefits they offer. While they stay with the matriarchal bands the immature bulls are subordinate to the adult cows in the group. One indication of their lower status is the lack of the aggressive vocalizations that older bulls exhibit during the breeding season. Due to their low social stature, they are not able to participate in the rut, even though many of them are physiologically capable of breeding.

Most cows in this group are able to breed, but in order to produce a calf at the age of two, these cows must have been bred as yearlings, and this is quite rare. Once they reach the age of two however, the majority of them, perhaps 50 to 80 percent, will conceive during the breeding season. Whether or not they conceive depends partially on body weight: larger cows are able to breed, whereas smaller cows are not. These young cows can be fairly domineering and have been observed sparring with two-year-old bulls.

A two-year-old plains bison cow. Her horns have begun to curve inward at the tip, but her bonnet is still undeveloped.

Male Two-Year-Olds

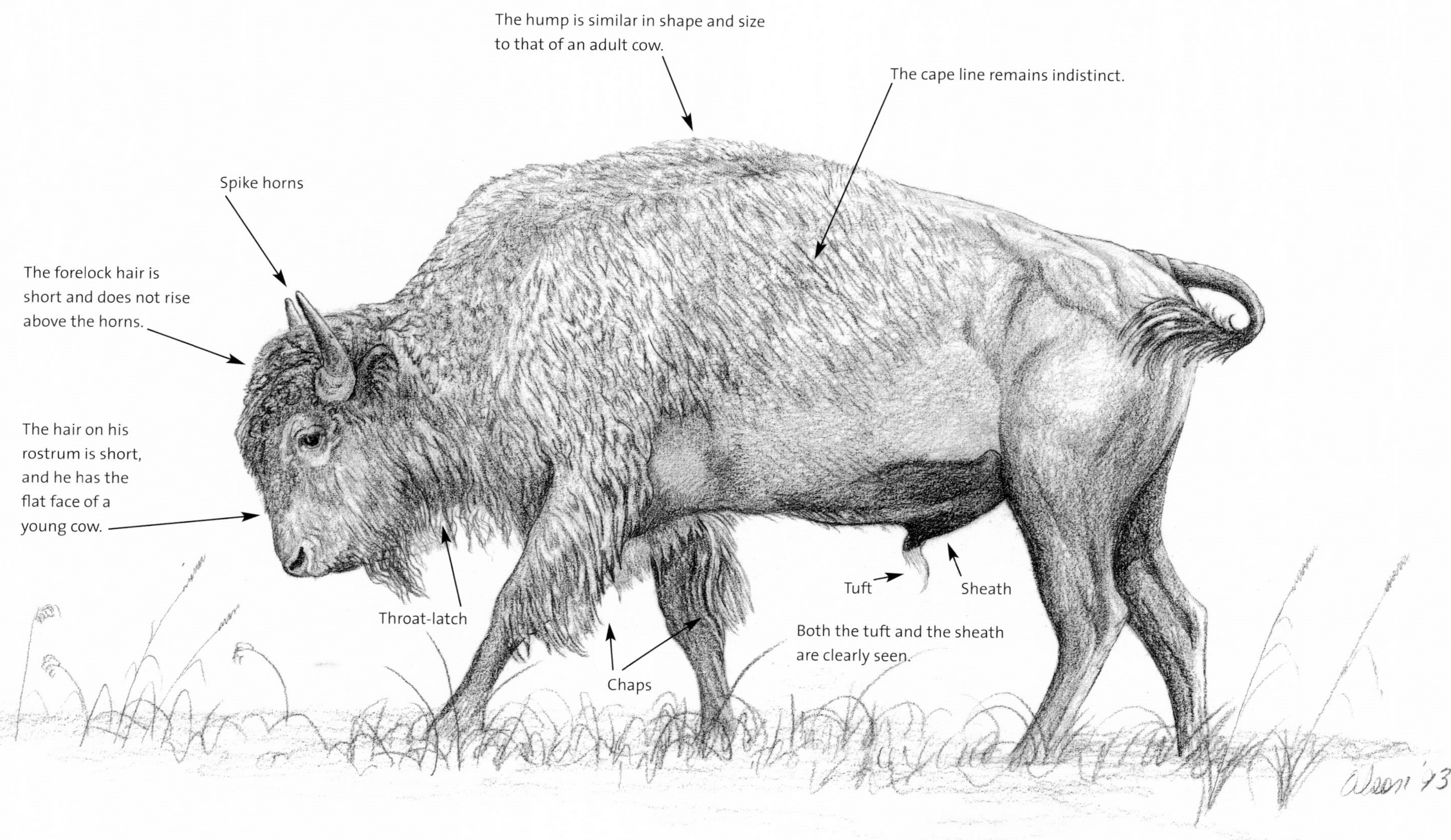

The two-year-old bull, while physiologically capable of breeding, is usually not permitted to do so by the older, more dominant bulls. Perhaps as a self-defense mechanism against aggression by these bulls, the two-year-old continues to wear a coat that resembles a cow's. The hair between his horns remains short and does not project above his horn tips. The hair on his face is short as well, and in profile his face looks quite flat. By this age, his horns project out and straight up, giving rise to the term "spike bull."

This young bull is slightly smaller than an adult cow and easily distinguished from her by the presence of his tuft, and by the size and shape of his horns. His chaps are becoming pendulous and are longer than the leg is wide. The beard and neck mane are larger than on a yearling, but the throat latch is still higher than his chin. The neck mane passes between his front legs on a level equal to the underside of his chest.

Female Two-Year-Olds

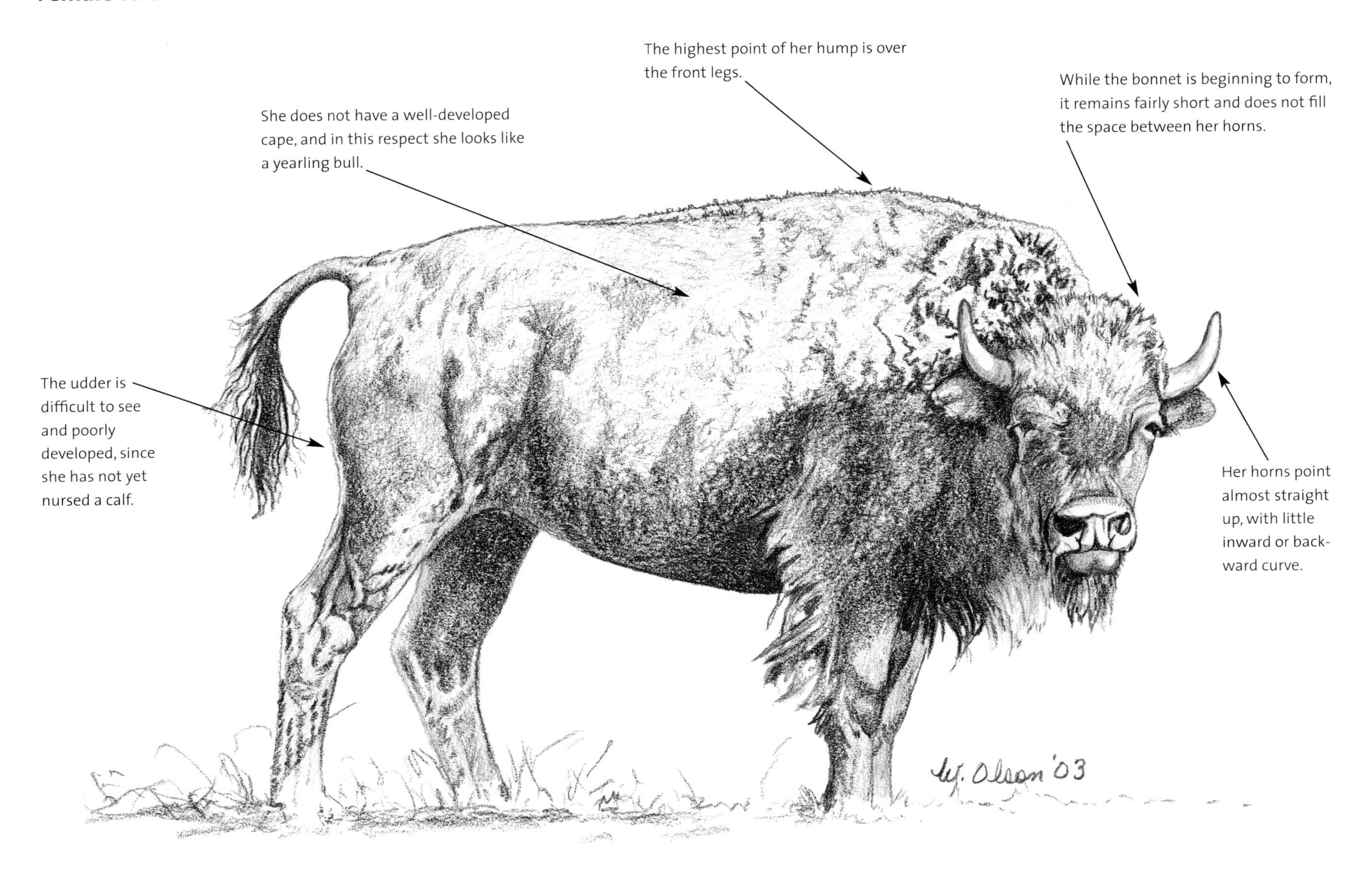

This young cow is entering her first reproductive year. If two-year-old cows enter the breeding season, during July and August, in good physical condition, most of them will conceive and nine months later produce their first calves at the age of three.

Two-year-olds cows can be difficult to distinguish from large yearling bulls, or smaller two-year-old bulls. The absence of the penis sheath is the primary indicator of her sex, and the size and shape of her horns, combined with her poorly developed cape and bonnet, are the principle indicators of her age.

Young Adults (36 to 48 months)

Social Status and Behaviour

By the time bison of both sexes reach age three, they have begun to assume the physical traits typical of their type. The males are easily capable of breeding receptive cows—physically, at least—but the dominant bulls still do not permit such rash impertinence. Young adult bulls will hang around the periphery of the herd, darting in to check a cow whenever the herd bull is preoccupied with a valid contender or so enraptured with a particularly appealing cow that he temporarily loses track of the youngster. The three-year-olds still like to play the part, though, and you can watch them smelling cows and lip curling over their delicious aromas.

Sparring matches are frequent, though not particularly impressive. These young bulls resemble teenage human males in many ways, being somewhat annoying to their older herd mates and downright obnoxious to younger males. While the actions of the three-year-old bulls within a rutting herd do not have a major impact on any of the herd members, they do train the youngster to be able to defend himself. They are preparatory training sessions for the day when he becomes serious about challenging a herd bull. This is the last summer that these bulls will be an integral part of the matriarchal bands, and following the rut most of them will drift away with the older bulls to form their winter bachelor groups.

The third year also marks a major change in the life of the females, for in the wild this is the age at which most of them will produce their first calf. Fewer cows in this group produce a calf than in the older age groups, and these first-time calvers may not be as good mothers as the older cows. When calves nurse, they do so until their appetite is satisfied and the calf breaks off the nursing bout, or the cow decides she has had enough and ends the bout. Young cows with their first calf tend to not let the calf nurse for as long or as frequently as the older cows do, and it is usually the cow, not the calf, that ends the nursing bout. Older cows let the calf nurse until it is full and usually wait for the calf to end the nursing bout and walk away.

Most parks and reserves must gather their bison herds off of the open range, into a series of pens or corrals. This is done periodically to check the health of the herd. These roundups can be very chaotic as the handling and sorting of the herd takes place. In Elk Island National Park, it is not unusual for two or three cows of the same age to come through together. They stick together through all of the sorting, mixing and chaotic movement of animals, probably because of their life-long friendship. These young cows may stay with their group within the cow-calf herd for the rest of their lives.

It can be difficult to distinguish three- and four-year-olds of either sex in the field. Some calves are born early, some late, often with up to six months between the first and last calves of the season. When these reach the age of three, there can be considerable differences in horn shape and size, body size, and so on.

This three-year-old bull is typical of his age group. This is the first year that he has attained the pelage characteristics typical of bulls and that he is physiologically capable of breeding. Bulls of all ages are impelled to investigate cows in heat, and during July and August a cow's urine is constantly tested to determine how receptive she is to being bred. A young bull might get away with testing a cow, as this one is doing here, but if she were really close to being receptive, neither she nor the breeding age bulls in the group would consider allowing him to get close enough to find out (right).

Young Adult Males

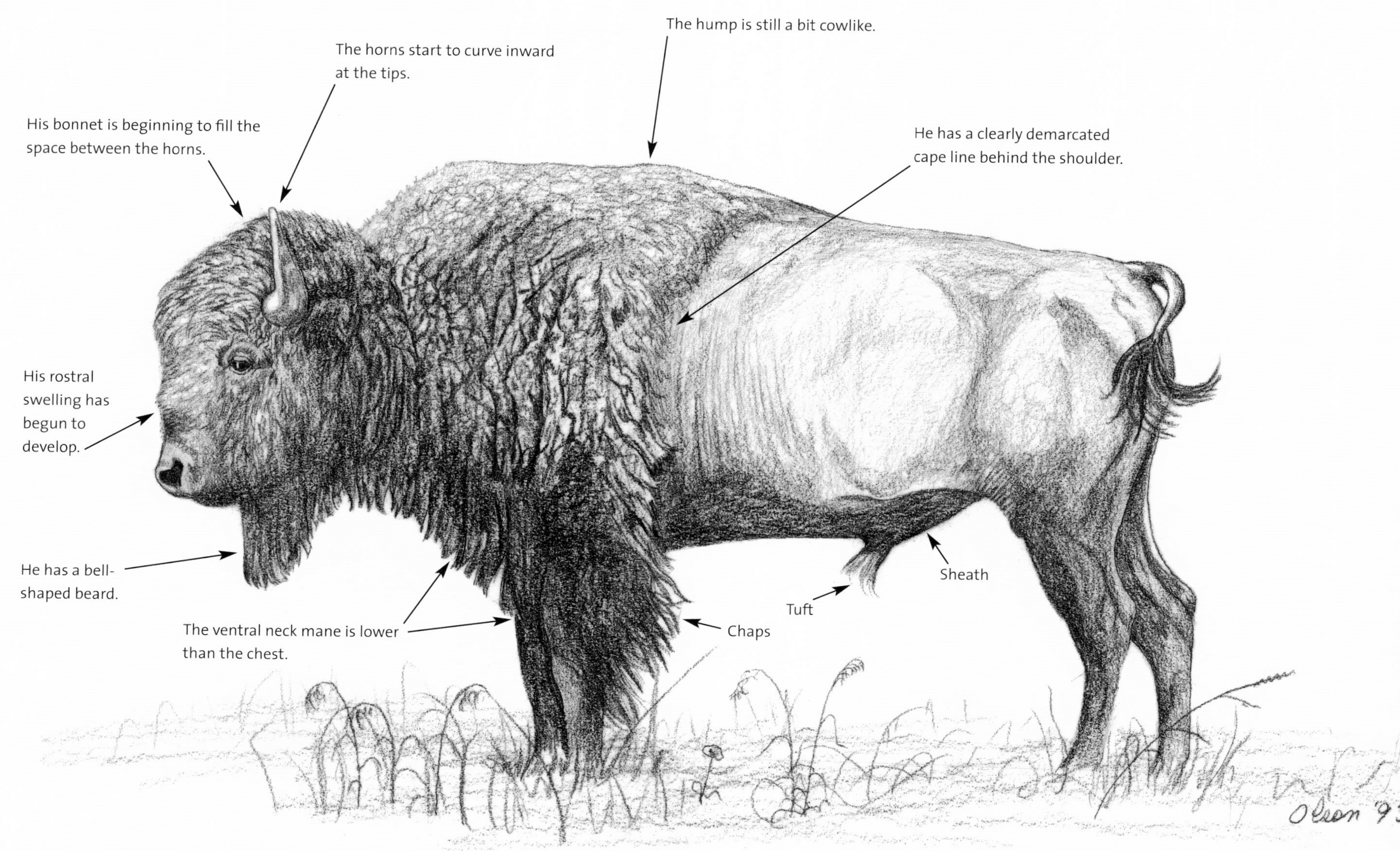

For the first time this young bull has shed his winter's coat and attained the appearance of an adult. The rear edge of his cape is clearly demarcated, with little or no hair left on the rib cage. His horn tips arch upward and show a definite inward curve at the tip. The bonnet between his horns is full and equal to the tip of his horns. The rostral swelling on the bridge of his nose is now noticeable, and his beard is assuming its bell-like shape. Finally, at the age of three, the plains bison bull has begun to assume the characteristics that help to define the subspecies.

He is clearly larger than the adult cows. At this age he is constantly testing himself—and his place within the summer hierarchy. Three-year-olds are fully capable of breeding, and while they know it, they also know they are no match for the older males; when a dominant bull approaches, these young bulls quickly make way for him. His pelage characteristics reflect this ascension to adulthood, with lush chaps and beard, a neck mane lower than the chest, and a well-developed tuft. His sheath, however, remains poorly developed. The key characteristic is his horns, which point straight up, with a little inward curve at the tip.

Young Adult Females

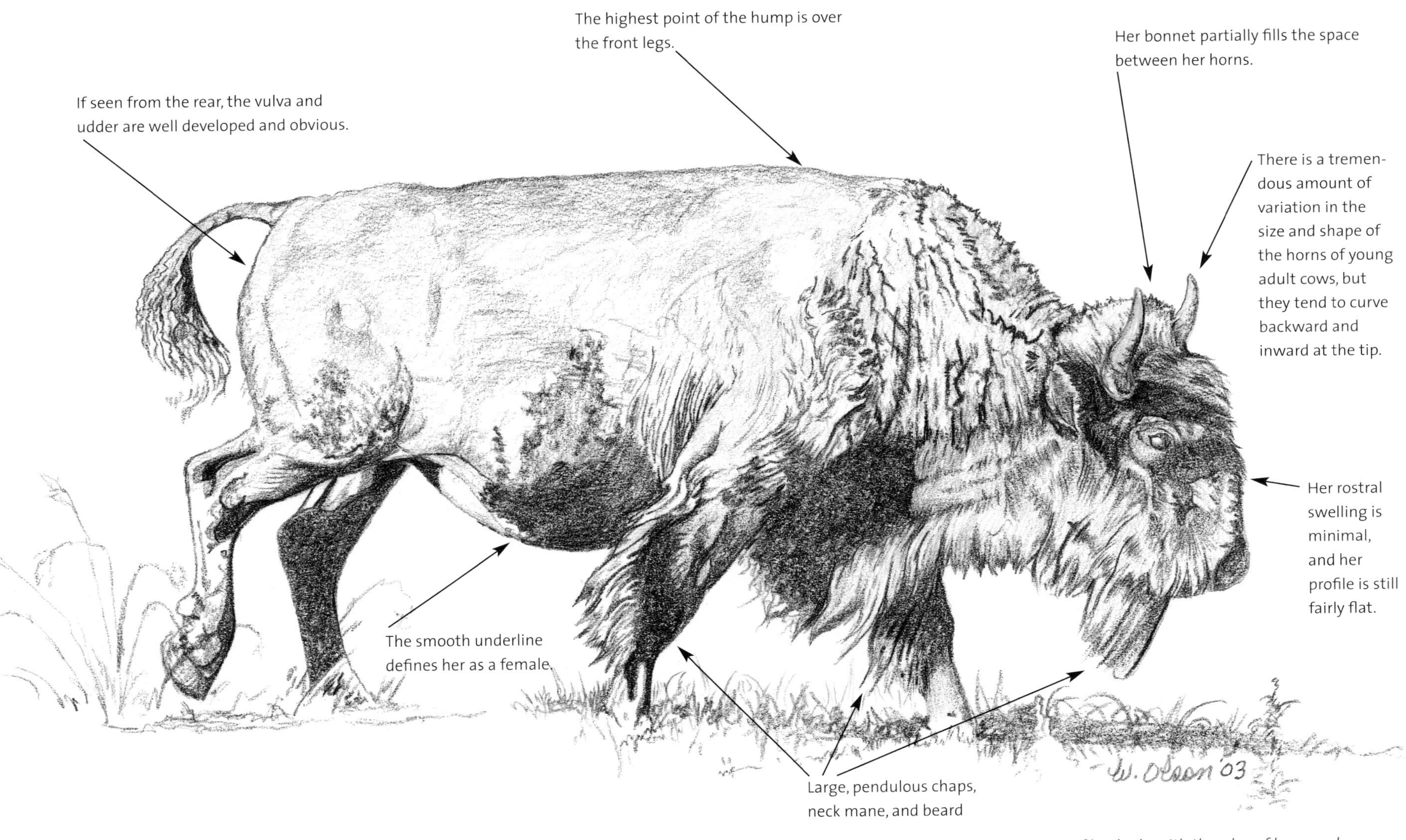

Spring arrives and with it, after a nine-and-a-half-month gestation, the birth of a new calf for this young mother. Bred at age two, she has now officially reached adulthood—and she looks the part. Her horns are well developed and gently curve up and inward. She has a bonnet of hair that partially fills the space between her horns, but the bridge of her nose is still flat. Her hump is gently rounded and balanced over her front legs. The cape on this young cow is lighter in colour than the rest of her body, with the edge of her cape less crisply demarcated than on a three-year-old bull.

In the field, she can be identified as a female by her thin horns, the lack of a tuft, and external genitalia under her tail. If she is nursing a calf, her teats and udder will be easily seen.

Through the process known as exfoliation, the upper tip of the bull's horn develops a frayed ring around its upper third. This happens as the horn grows, as it is abraded by ripped sod and battles with other bulls. The ring forms when the bull reaches about eight years of age, and it is the determining characteristic between the categories of "mature" and "dominant" bulls (left).

This mature bull, at the National Bison Range, Montana, is investigating a cow to determine whether or not she is in heat and ready to be bred. He is rolling back his upper lip in a behaviour known as "flemen." This allows him to roll the smell through his nasal passages to detect the levels of her hormones and thus whether she may be receptive to his further advances. While he is very capable of breeding her and will if permitted, the advance of an older dominant male will quickly send him packing. The absence of a ring around the upper tip of his horn places him in the mature bull category.

Directly behind him is a yearling male and above him, a yearling female. The difference in horn size and shape between males and females can be seen with these two yearlings (above right).

Bison cows have four teats, which supply highly nutritious milk to the voracious offspring. Calves usually nurse with their rumps against mom's shoulder and bunt the udder hard with their noses to start the flow of milk. By the time the calf is old enough to wean, its bunting almost lifts the cow off her feet (right).

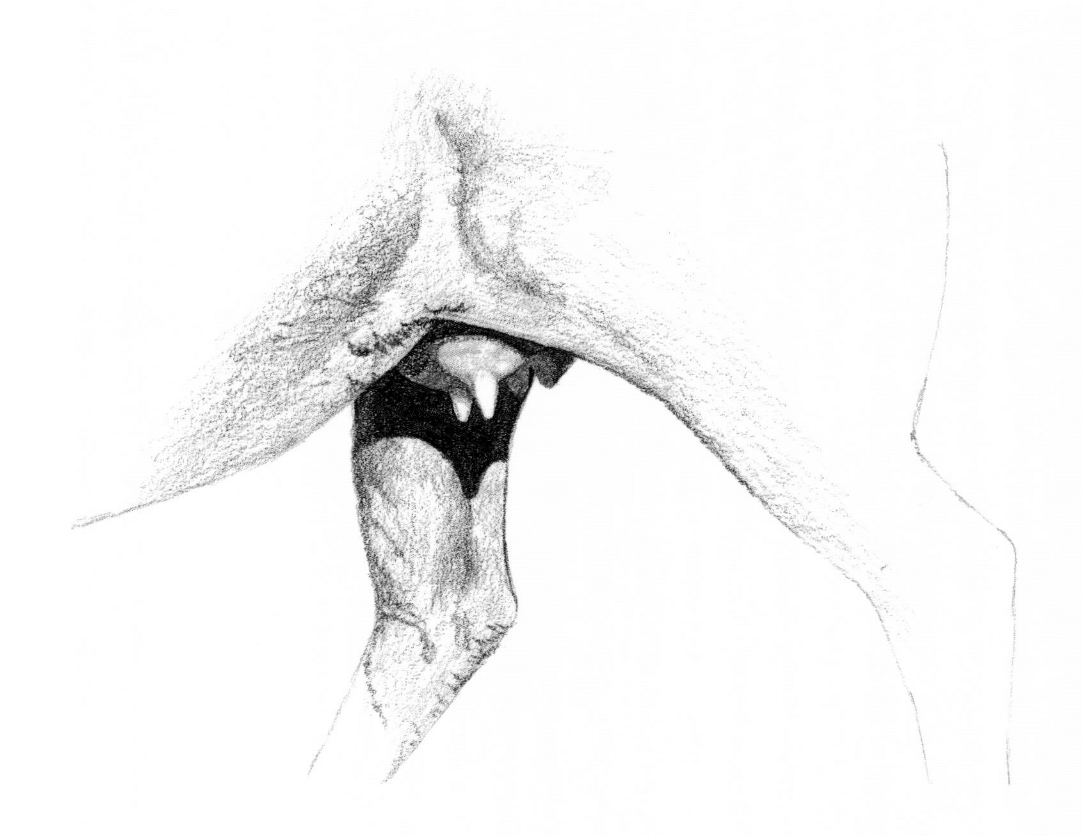

Mature Adults (4 to 7 years)

Social Status and Behaviour

Once an immature bull leaves the matriarchal band, he discovers that there are two principle groups of bulls. The first of these is the *mature bull* group. Plains bison in this age group are all physiologically capable of breeding and will do so in the absence of an older, more dominant bull. The primary distinguishing characteristic between a *mature bull* and a *dominant bull* is the lack of the ring around the tip of the horn, which usually forms at around age eight. While mature bulls without a horn ring will certainly breed if given the opportunity, bulls with the horn ring are more likely to do so, given their older age and greater experience.

As the bulls advance in size and age through the mature bull group and into the dominant bull group, they become more aggressive toward their rivals and the fights that occur begin to take on a very serious intensity. Bullering—the practice of one bull mounting another bull—is quite common in the younger ages, but as they advance in age this dominance behaviour declines in frequency and is replaced with more aggressive head-to-head battles. Subtle physical differences in appearance exist between each age of bull within this group, but they are difficult to detect and from the viewpoint of bison behaviour are insignificant.

Once cows reach age four or five, it is difficult to determine their age class. All cows tend to be equals until they achieve old age, and the majority of any bison population will be adult cows ranging in age from four to fifteen years. This is the most productive component of the herd with some of these cows producing a calf every year. Their ability to conceive and carry a calf to full term depends on their ability to gain weight each spring, and this in turn is reflected by the environments in which they live. Cows must attain an optimal weight following the calving season in order to be physically fit enough to conceive. Given good range conditions in the spring, they may produce a calf each year. Should they experience a severe winter or be injured or debilitated in any way and thus unable to reach their optimal weight, they will often skip a year and spend one season recuperating. The previous year's calf may benefit from the cow's inability to conceive, as it will continue to share the protection of its mother through into its second summer—something its peers may lack.

Mature Males
(approximately 4–7 years old)

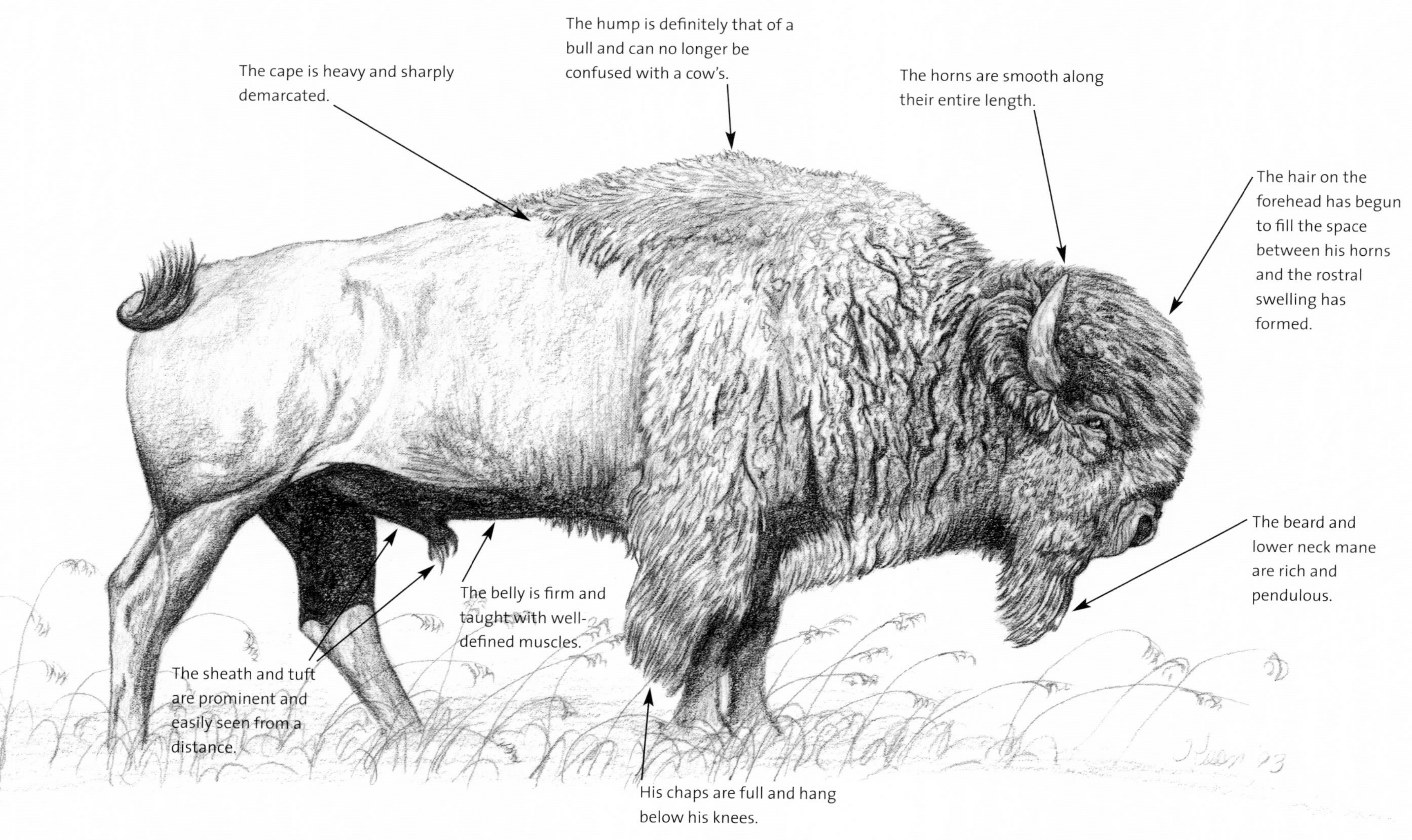

Mature bulls, such as this one, are the challengers in a rutting aggregation. There are two principle roles for bulls during the rut. The first and primary goal of all bulls is to be a "tending" bull—one that has a receptive mate and is guarding her against the advances of rival bulls. The second role is the "challenger" bull—one that does not yet have a mate and is looking to displace a tending bull. These challengers are large, aggressive males striving to attain ascendancy in the rut, and they can often be seen approaching a bull that is tending a cow. If his overt behaviour is strong enough to bluff the tending bull, the latter may yield the field to the challenger. If the tending bull is larger, older, more experienced, or simply more determined, the challenger may try the visual bluff a bit longer, then stalk off in search of a less dominant bull to harass.

The primary visual differences between these mature bulls and their older counterparts (the dominant bulls) is the lack of the horn ring and the presence of a firm, flat stomach. They may have less hair on the bonnet, but their beards, chaps, and neck mane are all very well developed.

Mature Females
(4–14 years old)

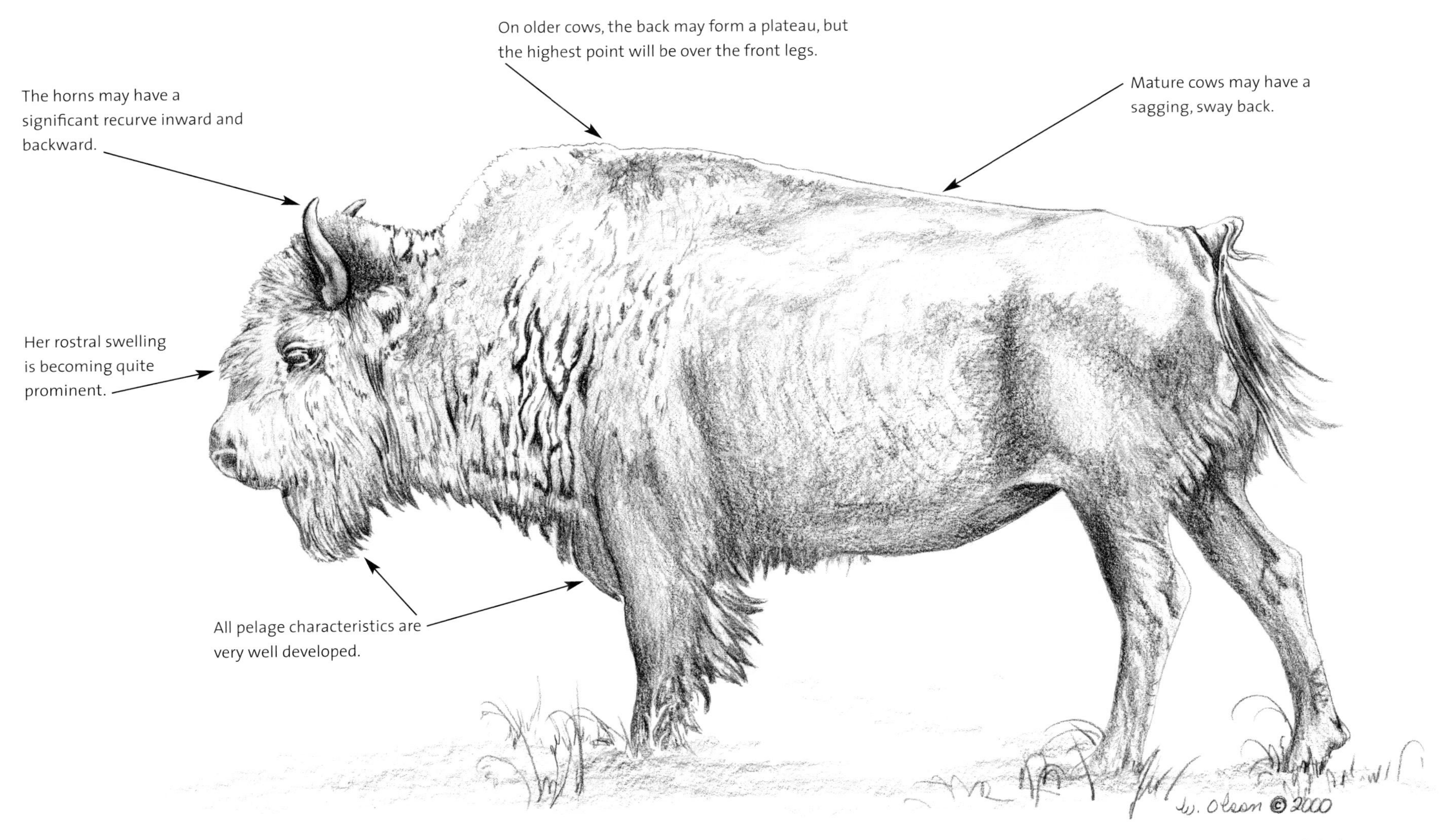

While it is possible, based upon horn and facial hair morphology, to separate cows into rough age classes beyond the age of four, there is no behavioural or biological reason to do so, as all of these cows contribute equally to bison society. As a result, the group comprised of cows from age four to approximately fourteen is represented by the cow shown here.

The photographs on pages 100–102 demonstrate the changes in the shape of the hair on the bridge of the nose as a cow ages, and this can be useful in assigning rough ages to adult cows. These photographs also show the general shapes of female bison horns as they age, and used with other characteristics, horn shape and size can be useful in placing cows into rough age groups.

Dominant Males (8 to 14 years)

Social Status and Behaviour

Mature bulls have struggled and fought their way up the social ladder, and by the time they reach the age of about eight years, they become dominant bulls. They are distinguished from their younger rivals by the presence of an exfoliated horn tip. These are the most majestic and intimidating bison on the plains of North America. Often weighing more than a tonne, they stride through bison society with utter impunity, the lesser bison moving aside like waves under the prow of a charging ship. Nobody gets in their way or stays there for very long.

The most serious battles for herd dominance and fights over receptive cows are conducted by these scarred veterans. This is the age of male that suffers the highest mortality rate, for these summer battles exact a deadly toll on the losers. Often a challenger bull arrives at a herd of bison cows and calves being tended by one of these dominant bulls, and there will then take place a choreographed series of poses and posturing by both bulls to determine who is dominant and who is not. Most of these encounters are resolved quickly without physical contact. But should the contender be close in age, size, or dominance, the level of interaction will escalate from dominance displays to physical contest—often brief, with the tending bull quickly winning or losing to the contender. And should the contender be an equal match to the tending bull, then the battle begins. It may last anywhere from a few minutes to as long as an hour. These battles can be very intense and nasty, and they can result in severe injuries to both parties. During July and August, lone breeding-age males are alone either because they have done all of the breeding they want to and have voluntarily left the herd or because they have been thrashed so soundly that they have retreated to a solitary life as a means of survival.

Dominant males roam the herd seeking a cow that is close to being in heat, and once they find her, they spend all of their time tending her and protecting her from the advances of other bulls. This tending behaviour requires considerable energy reserves, and the bulls that breed the most cows are those that have stored fat reserves that allow them to stop grazing for long periods of time. If they have low energy reserves, they may tend and breed only three or four cows before they have to leave the rutting herd and look for forage to rebuild their strength. To stay with the herd would mean putting up with the constant challenges by younger bulls wanting to test the vigour of the tired and worn out dominant bull, and by other dominant bulls who are striving to maintain their status. It is simply easier to move off to the side and get some rest. If they regain their strength early enough in the summer, they may re-enter the herd later in search of more females to breed.

Wounds such as this are a common sight on breeding-age bulls, and while they have the potential to be life-threatening, usually they heal fairly well (top left).

The rutting herd is a scene of constant turmoil as bulls enter the herd, get beat up, and depart. Bulls possessively tend receptive cows, calves scurry to keep out of the way, young bulls try in vain to ascend the social totem pole. Subtle and obvious dominance battles are underway between the older bulls.

In this photo, two bulls of roughly equal dominance have begun to communicate their intentions to each other. The bull grazing is being approached by the contender bull. While he seems to ignore the contender, he is in fact paying very strict attention to him. The contender is exhibiting physical aggression: he is roaring loudly and swinging his head from side to side as he approaches, in an attempt to bluff the tending bull into believing that he is the more dominant of the two. If he is successful, the tending bull will turn and walk away. If he is not, an intense head-to-head battle may quickly ensue (right).

A dominant bull will aggressively approach a wallow and displace whoever is in it. Then he rips at the ground in a flurry with his front feet, his horns, or both to loosen the dust. He urinates into the dust and proceeds to thrash and roll about in this messy concoction. This mud bath serves to identify the wallow as his territory, places his breeding smells for others to smell and respect, and makes him more attractive to receptive cows. This rutting behaviour at the wallow is often a prelude to an aggressive approach to another bull.

With tail and nostrils flared, this plains bison bull investigates an estrous cow that appears more interested in a good scratch than in courting.

Dominant Males
(approximately 8–14 years old)

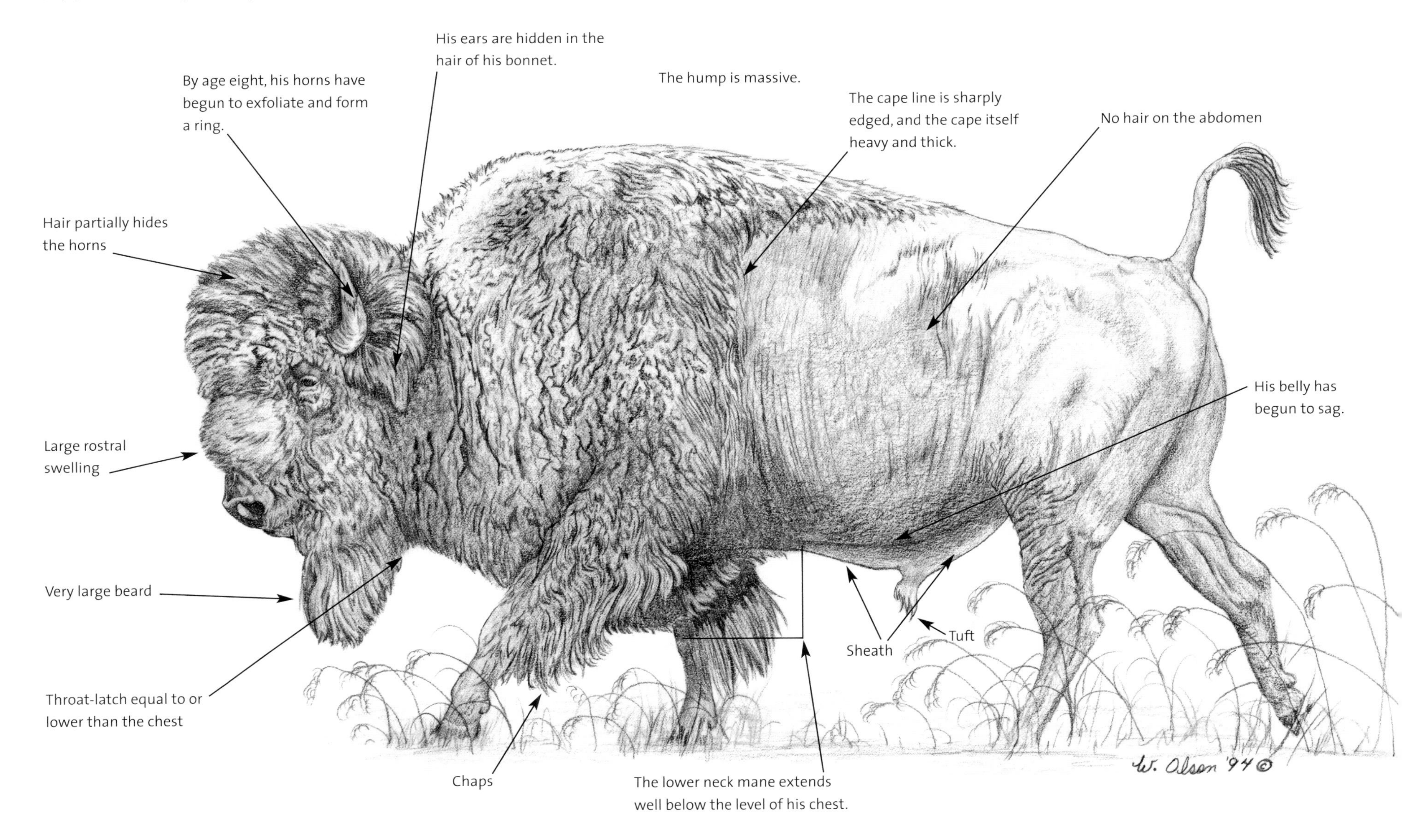

The dominant breeding bull is an animal that exudes vitality and raw power. Because of his sheer size and aggressive nature, this is the bull that resides at the top of the bison social structure. Even some of the older bulls give way to him because of his aggressive attitude. When a bull like this arrives at a herd during the breeding season, you can often hear him chuffing and roaring to let all contenders know that he has just entered the arena.

By this age, his stomach has begun the inevitable downward sag typical of advanced middle age. All of his pelage characteristics are magnificent and well developed. When he strides purposefully toward another male or an estrous female, the beard and chaps swing and sway with each and every step.

Old bulls like this one, in Elk Island National Park, often leave the company of other bison, preferring to be off by themselves or possibly with another bull of similar age.

The Aged
(15 years +)

Social Status and Behaviour

In most bison populations, reaching old age is a significant achievement. The risks associated with living in an inhospitable environment make it possible for only the very fittest to survive to the ripe old age of fifteen. Males especially are unlikely to live long lives due to their desire to explore and establish new ranges, to their life-and-death battles during their prime breeding years, and to the fact that as an older adult they often live alone and forego the safety that large herds provide. These old bulls have given up on participating in the rut by the time they are about twelve to fifteen. They live out the remainder of their days in groups of two or three, or often alone, occupying habitats of lower quality than those occupied by the cow-calf herds, and unless a debilitating disease, injury, or accident occurs, they can live long and uneventful lives. Plains bison bulls in Elk Island National Park have been known to live into their early twenties.

Old cows are the matriarchs of the herd. They may have several generations of offspring surrounding them in maternal herds, and they probably provide a knowledge base upon which other herd members rely. Some research has shown that bison herds are composed of randomly associating individuals with no clear structuring by family lines. In most herds, though, regardless of whether they are composed of family groups or randomly associated individuals, some groups are sure to have one or two matriarchs as guides and mentors for the younger cohorts.

During the periodic bison roundups conducted by the park managers to ensure that the herds are healthy, all bison receive a small ear tag to enable their identification and of all other herd members. In Elk Island National Park, it is not uncommon for older cows, in their late teens and early twenties, to come through the handling operation together; and they do it year after year. This is another example of the life-long friendship forged as yearlings following their abandonment by their mothers. Year after year these old friends make it through another winter together.

Bison are one of the longest-lived large mammal species in North America, and in Elk Island National Park, Alberta, cows as old as twenty-six have been recorded. This old girl is a typical example of an aged cow. She has lost her right horn sheath, has a large bonnet and a mop of hair on the bridge of her nose.

Aged Males
(15+ years)

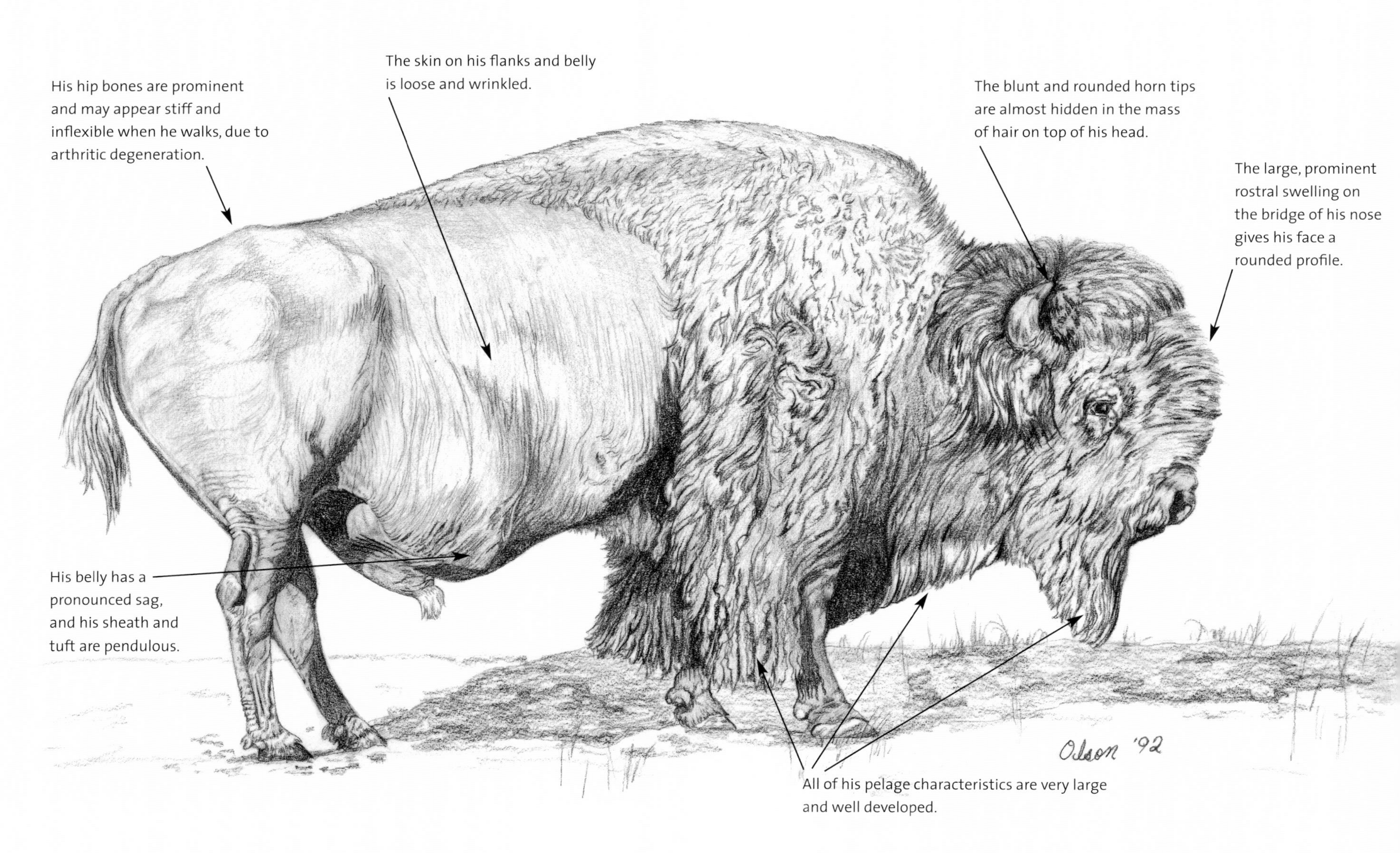

This is a classic example of a bull that has reached old age. His belly has lost the battle with gravity and time. He bears the visual evidence of a long and hard-fought life, with rips and scars and hidden wounds. Arthritis can be the bane of his existence, and an examination of an aged bull's skeleton often reveals massive degeneration of the hip joints. The hocks are swollen and stiff, and fast movements are merely a memory.

Not much bothers these old-timers—not cold, not heat, nor impertinent youngsters seeking advancement up the bison social ladder. They just do not care anymore. Their only apparent goals in life are to be left alone and to find a nice patch of fresh grass to eat and a shady wallow to lie in while they ruminate upon life.

Aged Females
(15+ years)

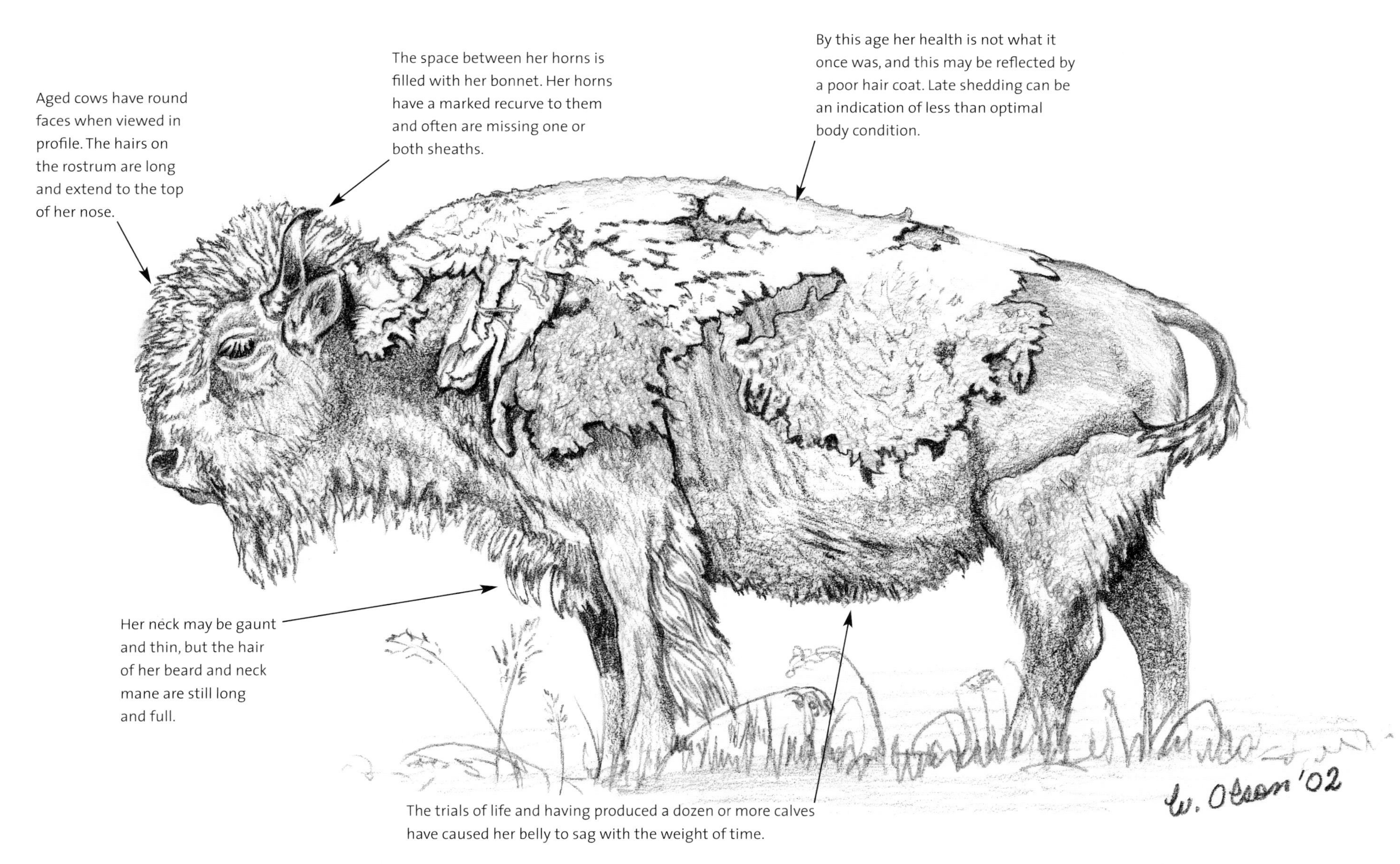

In wild populations where bison are exposed to all of the risks associated with living in harsh environments, the ability to live to old age is rare and the opportunity to observe a cow in this age group is equally uncommon. These are the matriarchs, the great-great-grandmothers of bison society. Through the course of their lives, they could easily have produced a dozen calves by the time they reach age twenty. The life experience gained by these cows is passed on to their offspring in many ways. The locations of secluded calving areas, trails to water holes, mountain passes, and seasonal migration routes were learned by this cow when she was young, and she has passed them on to the generations that followed her. This passing on of knowledge is part of a continuous stream from one bison to the next that has been going on, uninterrupted, for as long as there have been bison.

This cow became mired in the mud along a receding lakeshore and eventually died there. She is recognizable as a female because the base of her horn is smaller than her eye socket.

The Dead

Eventually the time arrives when life must come to an end, though bison are one of the longest-lived North American mammals. Bison living in wild places may succumb to a variety of traumatic events, early in their lives or as old-timers that have lived long and contributed to society through their accumulated knowledge and the continuation of their genes.

In the course of my duties as a park warden and in my private life as a bison rancher, I have witnessed the death of many bison. With the exception of old loner bulls, death is seldom, if ever, a solitary affair. Many times I have seen a dying animal tended by his or her herd mates, which seem to try to encourage the animal in distress to get up, to stop this irrational behaviour, to come along with the rest of the herd. I have watched them gently hook the downed animal with their horns, nudge it with their noses, and paw at it with their hooves. I have watched mothers stay with a dead newborn for several days, and I have seen calves stay with their deceased mothers long after the rest of the herd has left the area. In most parks and preserves, the dead remain where they lie, at least for a short while. Then they begin a new journey, a new migration across the land, as scavengers and predators alike—from bears and wolves to dung beetles and flies—take their portion of the banquet off to their respective dining areas.

Travellers to the backcountry of bison refuges may be lucky enough to stumble across the remains of a winter- or predator-killed bison and have the opportunity to examine the animal from a different perspective. If you have the patience to examine a carcass in detail, it is often possible to determine why the animal died, how old it was when it died, and what sex it was. Appendix 1 will aid you in deciphering the clues left behind to transform a pile of weathered and tooth-gnawed bone into a life history of the animal that was.

It is illegal in most protected areas to remove bones, skulls, or other remains of bison, or from any other animal. There is also an ecological reason: the bones left behind provide a rich and long-lasting reserve of calcium and other trace minerals for rodents and other members of the ecological community shared by bison. So leave them where you find them for the next hiker and for the next mouse.

Afterword

It is with considerable hope and optimism that I finish this book—hope that the information provided here makes your next visit to a bison refuge a safe one, both for you and for the bison you encounter; hope that you come away from this book enriched and enlightened.

If even one person walks away uninjured from a close encounter with a bison as a result of knowledge gained here, or if this book helps someone to understand the composition of a herd that he or she observes, then my goals have been met. With the continued protection of this magnificent species in the refuges of North America, the travelling public will be able to return to camp at the end of the day, and say to their neighbours

"I saw a buffalo today, and it was a good day."

Appendix I
The Bare Bones of It: How to Determine the Age and Sex of a Bison Skeleton

Why would anyone want to gather information about the skeleton of a bison? There are various reasons. Poachers are active in many protected areas, and it helps the investigating officers if they are able to determine the sex and relative age of the deceased animal. Wildlife biologists want to be able to monitor mortality rates by age and sex so that they can more accurately predict population change over time. And for others, it is simply fun to be able to investigate and solve the riddle of who the animal was.

Is It a Bison?

Although uncommon, in places bison and cattle coexist on the same landscape. In this case, or if you are unsure whether the bones in question are indeed bison, examine the placement of the *zygomatic arch* (see photograph on page 85) in relation to the top of the skull. In bison, the top of the skull obscures the arch; in cattle, it is visible from above.

Determination of Sex

Once you have determined that the bones in question are those of a bison, the next step in the investigation is to determine its sex. If the carcass is intact and you see the external gentalia, then it is easy to determine whether it is male or female based upon the information presented in Chapter Four.

Should the skeleton be intact with the bones still joined together, the task remains fairly easy, but if all you have is a pile of scattered bones, the task becomes more challenging, yet more rewarding. In most cases where the bison has died of natural causes, the skull will be found within a short distance of the rest of the bones, even if scavengers have been at work. Having access to the skull makes sexing the carcass that much easier.

Sexing the Skull

Examine the size of the eye socket in relation to the size of the horn core for adult male and female plains bison. On males, the diameter of the base of the horn core is equal to or greater than the diameter of the opening of the eye. Females have much finer horns, with the horn core equal to or less than the diameter of the eye socket.

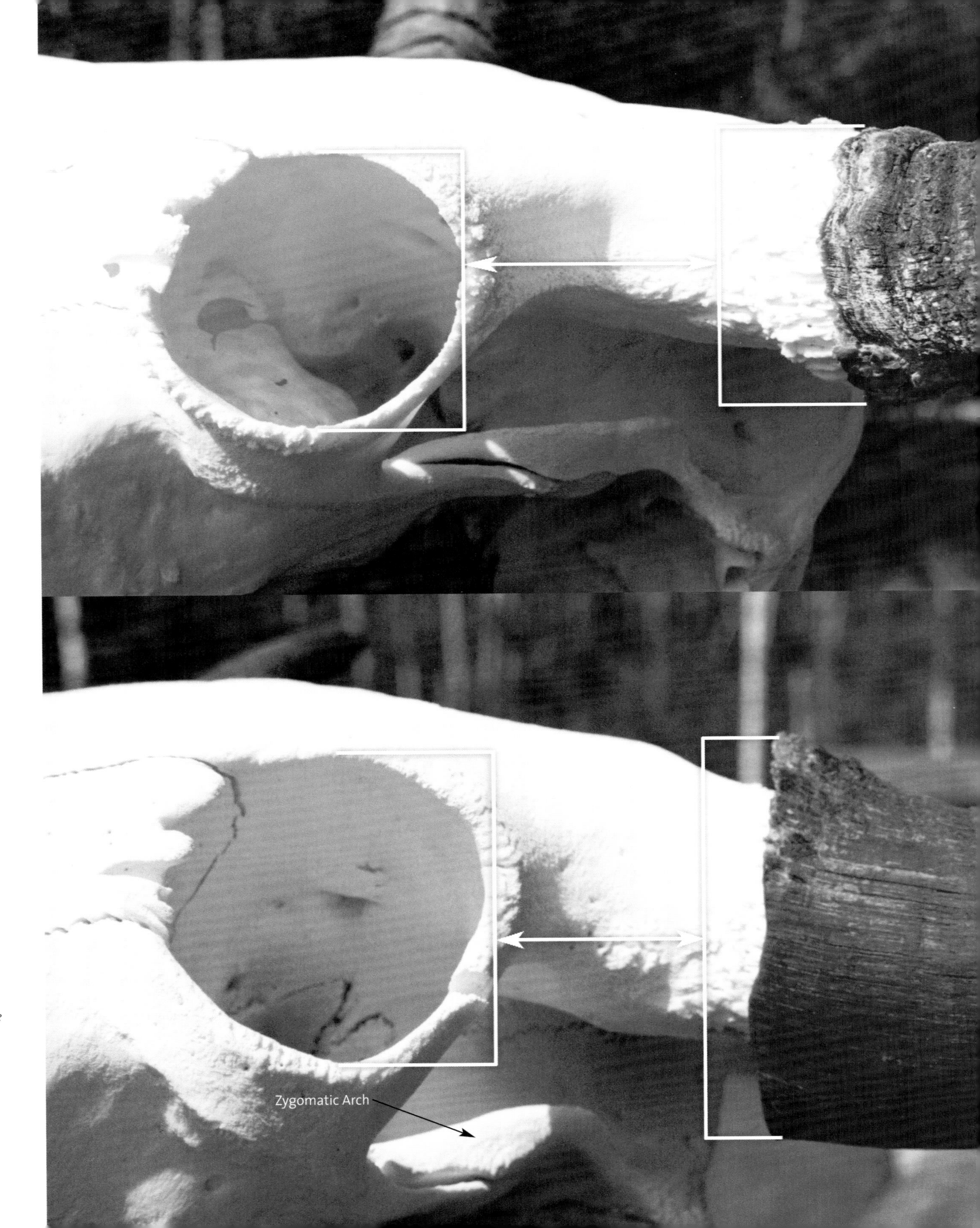

In the skull of an adult female plains bison, the horn core is equal to or smaller than the opening of the eye (above).

This skull is from a two-year-old bull. His horn sheath is larger than the opening of the eye, and this difference becomes even more pronounced as a bull ages (below).

Table A1.1 Measurements Used to Determine the Sex of Skulls from Adult Plains Bison

Measurement	*Male*	*Female*
WHCO	265 mm (246 to 289)	209 mm (190 to 222)
RW	119 mm (103 to 132)	99 mm (83 to 110)
GWA	256 mm (229 to 276)	205 mm (189 to 220)
GPW	318 (282 to 352)	257 mm (237 to 275)

WHCO—Width of cranium between horn cores and orbits
RW—Rostral width at maxillary—premaxillary suture
GWA—Greatest width at auditory openings
GPW—Greatest post-orbital width

Data from van Zyll de Jong 1986.

Determination of Age

THERE ARE SEVERAL fairly accurate field methods of determining the age of a bison at death. The most accurate is the schedule of tooth replacement. Like humans, bison are born with deciduous teeth, baby teeth that serve us through our first few years, which are then replaced with permanent and more durable teeth.

An examination of Table A1.2 and the photographs on the following pages permit the investigator to determine age fairly accurately, up to five years, based upon tooth replacement. After age five, determination of age becomes more subjective, but bison can still be placed into age classes similar to those presented in Chapter 4.

Bison can be placed into various age groups based on the amount of wear experienced by the teeth as the animal ages. You need to examine the teeth in the lower jaw of the animal, so an attempt must be made to locate it. The jaw is usually separate from the skull and often is broken into two pieces. With skeletons that have been in the environment for a long time, it is rare to find the incisor and canine teeth, since they are poorly attached and often fall out. The molars and premolars, however, are well-rooted, and after the age of one year, can by themselves yield vital clues to assigning age.

Note that there are regional differences in the amount of wear on teeth. Bison living in areas with sandy soils will wear their teeth down faster than others do, and depending upon the month of birth, growth rate, and nutrition, there may be variations of several months around the timing of tooth eruption. As a general guide, though, the teeth provide clues to place bison into recognizable age classes. The age classes used for determining age by the characteristics of teeth, vary slightly from those used to determine the age by looking at the live bison. For the purposes of understanding the role of the individual in the population though, the information in the following table serves the purpose.

Table A1.2 Age Determination by Tooth Examination

Age of Animal	*Tooth*	*Details of Wear*
Young Adult (4–6 years)	C–1	wear not exceeding 2 mm
	M–2	style a circle or loop
	M–3	style not worn, or tip only worn
Adults (7–11 years)	C–1	wear between 2 and 4 mm
	M–3	style rarely a loop
Aged (12+ years)	C–1	wear more than 4 mm
	M–3	style usually a loop

C=Canine
M=Molar

Source: Fuller 1959.

Table A1.3 Bison Tooth Replacement Schedule

Age	*Incisors*			*Canine*	*Premolars*			*Molars*		
	I–1	I–2	I–3	C–1	P–2	P–3	P–4	M–1	M–2	M–3
1 year	D	D	D	D	D	D	D	P	(P)	
2 years	P	D	D	D	D	D	D	P	P	(P)
3 years	P	P	D	D	P	(P)	D	P	P	(P)
4 years	P	P	P	D	P	P	P	P	P	P
5 years	P	P	P	(P)	P	P	P	P	P	P

D=deciduous teeth
P=permanent teeth
(P)=permanent teeth that are in the process of erupting

Data from Fuller 1959; Giles 1971.

Explanation of Tooth Eruption

The incisor bar to the right shows I–1 as fully erupted, I–2 as deciduous but in the process of being replaced with a permanent tooth, and I–3 and C–1 as deciduous baby teeth. This tooth row is from a bison that was 2 1/2 years old when it died.

In the lower photo, all incisor teeth are now permanent. The right deciduous canine has fallen out and is being replaced with a permanent tooth. The left deciduous canine remains in place, with a just barely visible permanent tooth replacing it. This bison was 5 1/2 years old when it died.

I=Incisor
C=Canine

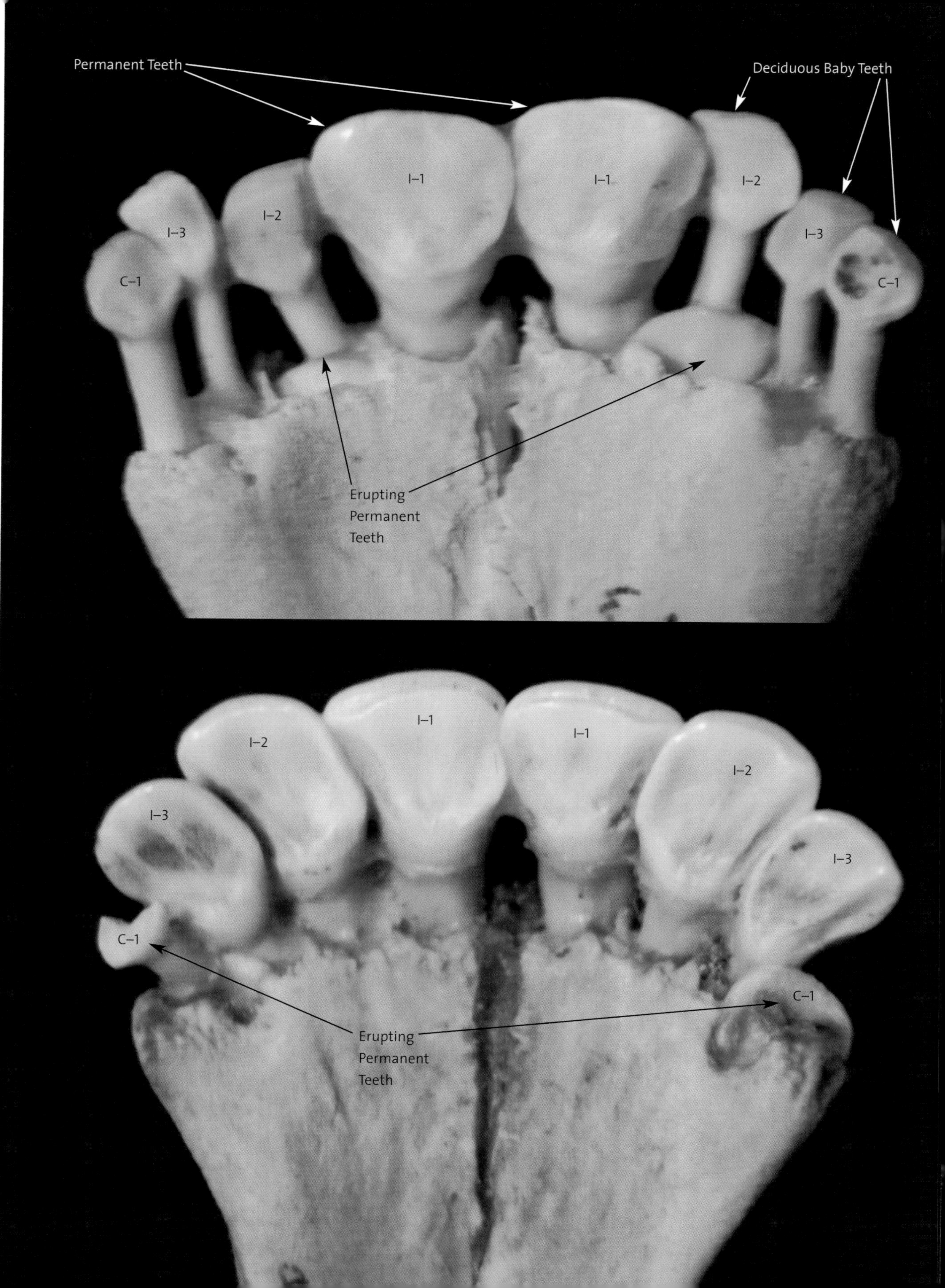

The Sequence of Eruption and Tooth Wear of Bison Teeth

Tooth replacement and wear patterns from Elk Island National Park plains bison. All samples are from plains bison during mid-winter. The occlusive surfaces (those areas of the teeth that experience wear from grinding action) are shown as darker lines of dentine, surrounded by light areas of enamel. As the tooth erupts and begins to experience use, the width of the wear lines increases. The row of teeth shown here is referred to as the incisor arcade. The loss of teeth, or deformities due to old age or excessive wear may make it so difficult for the bison to eat that they eventually die from malnutrition.

I=Incisor

Calf—6 months old
No wear on the occlusive surfaces of the incisors or canines. All teeth are deciduous.

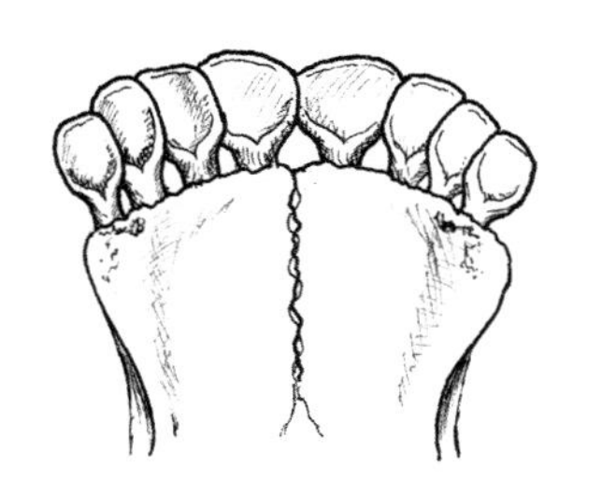

Yearling—1.5 year old
Slight wear on the occlusive surfaces of I–1 and I–2. All teeth are deciduous.

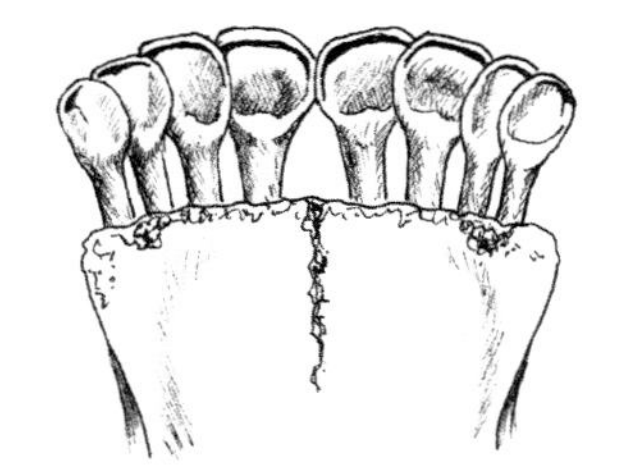

2.5 year olds
I–1 now permanent, all others are deciduous. No wear on the occlusive surface of I–1, moderate wear on I–2, I–3 and canines.

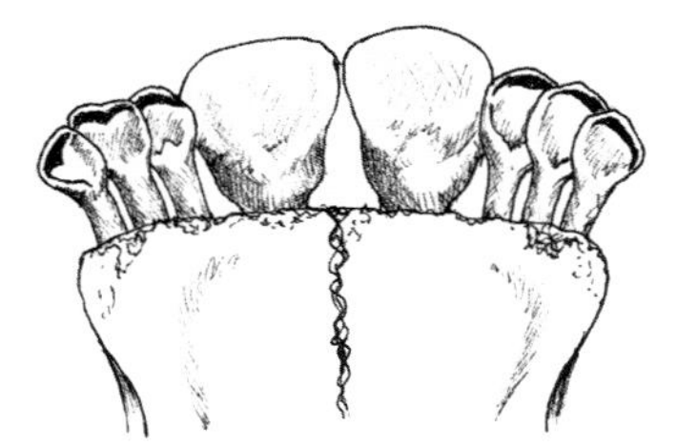

3.5 year olds
I–1 and I–2 now permanent, I–3 and the canines are still deciduous. Light wear on I-1.

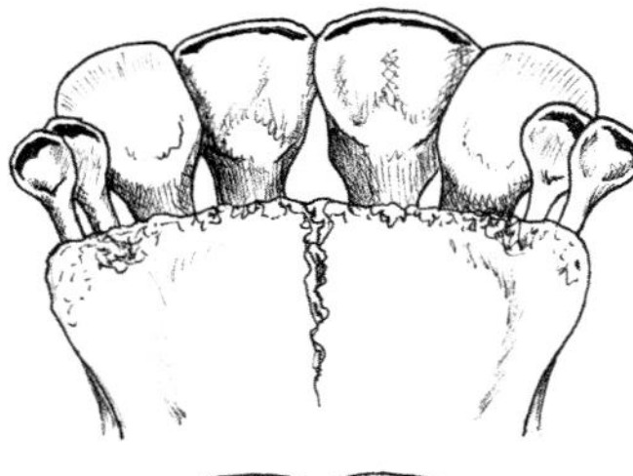

4.5 year olds
I–1, I–2 and I–3 now permanent. Light wear on I–1 and I–2, no wear on I–3; canines are still deciduous.

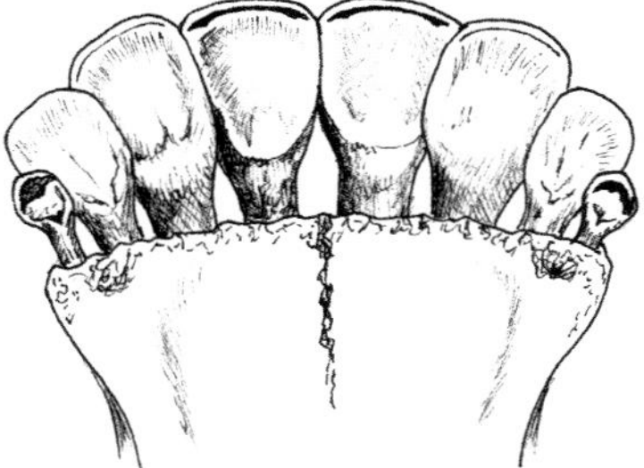

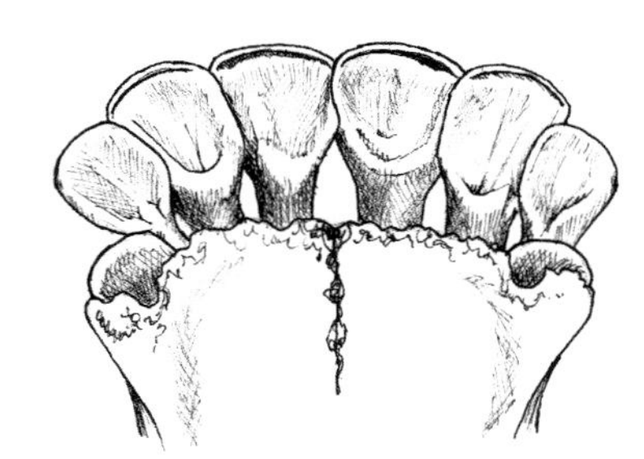

5.5 year olds
All incisors are now permanent and permanent canines are in the process of erupting. Light wear on all incisors.

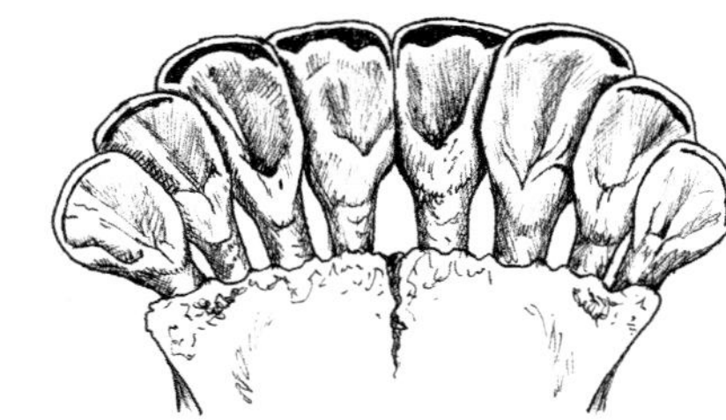

6.5 year olds
All teeth in the arcade are now permanent. Moderate wear on I–1, I–2 and light wear on I–3. Very light or no wear on the canines.

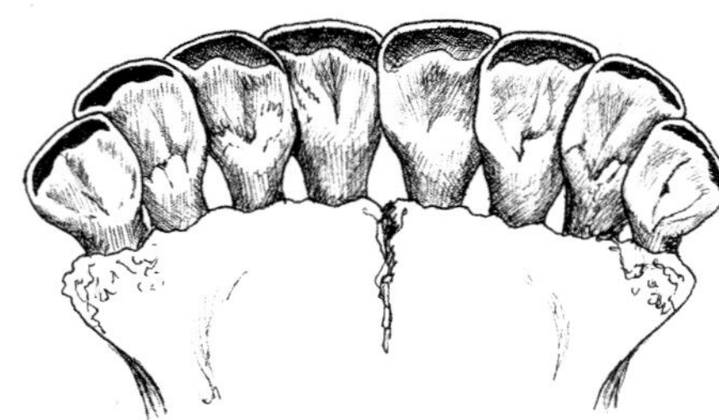

8.5 year olds
Moderate wear beginning to form on the canines.

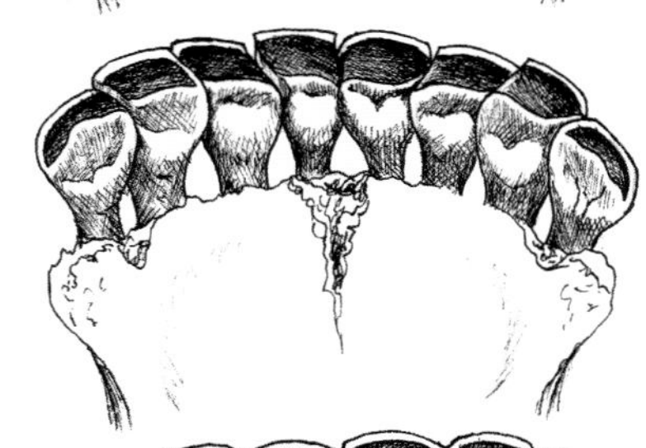

16 year olds
Moderate to heavy wear on all teeth; sharp edges and deformities may be evident.

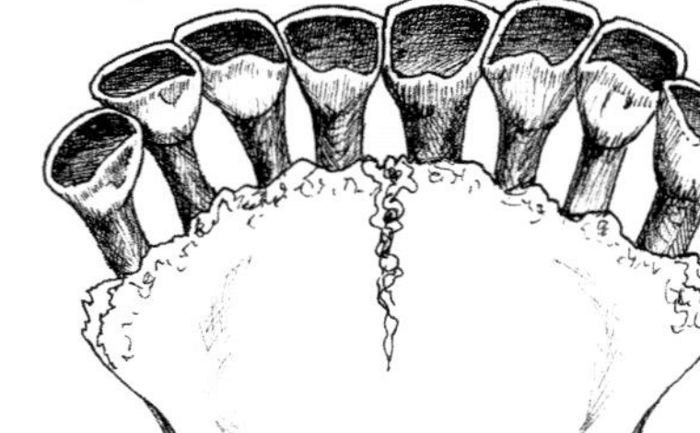

18 year olds
Heavy wear on all teeth; some teeth may be missing or misshapen.

A Dichotomous Key to Aging Bison by Teeth Characters

1. Deciduous teeth all present
    2. Second molar absent
        3. First molar not visible .... 3 mo.
        3. First molar erupting to fully grown .... 3–12 mo.
    2. Second molar erupting to fully grown....................................... 1 yr.
1. At least central incisors permanent
    4. **C–1** deciduous or shed, never permanent; **I–2** and **I–3** deciduous or permanent
        5. **I–3** deciduous; **I–2** deciduous or permanent; third cusp of **M–3** not in wear; style on **M–1** just beginning to wear, or worn to a circle (rarely a loop) ..... 2 yr.
        5. **I–3** permanent; **C–1** deciduous or shed; style on **M–1** typically a loop, occasionally a circle or not worn .... (in part) 3 yr.
    4. All teeth, including **C–1**, permanent
        6. No visible wear on **C–1**; style of **M–3** not in wear
            7. Typically no wear on **I–3** as well as **C–1**; style of **M–2** not in wear or with wear on tip only .... (in part) 3 yr.
            7. Typically slight wear on **I–3**; style of **M–2** usually a circle. Occasionally slight wear on **C–1** if style of **M–2** not in wear .... 4 yr.
        6. Visible wear on **C–1**
            8. Wear on **C–1** not exceeding 2 mm; **M–2** style a circle or loop; **M–3** style not in wear or with tip only worn................................ young adult
            8. Wear on **C–1** exceeding 2 mm
                9. Wear on **C–2** 2–4 mm; **M–3** style rarely a loop .... adult
                9. Wear on **C–1** more than 4 mm; **M–3** style usually a loop........... aged

C=Canine
I=Incisor
M=Molar

Source: Fuller 1959.

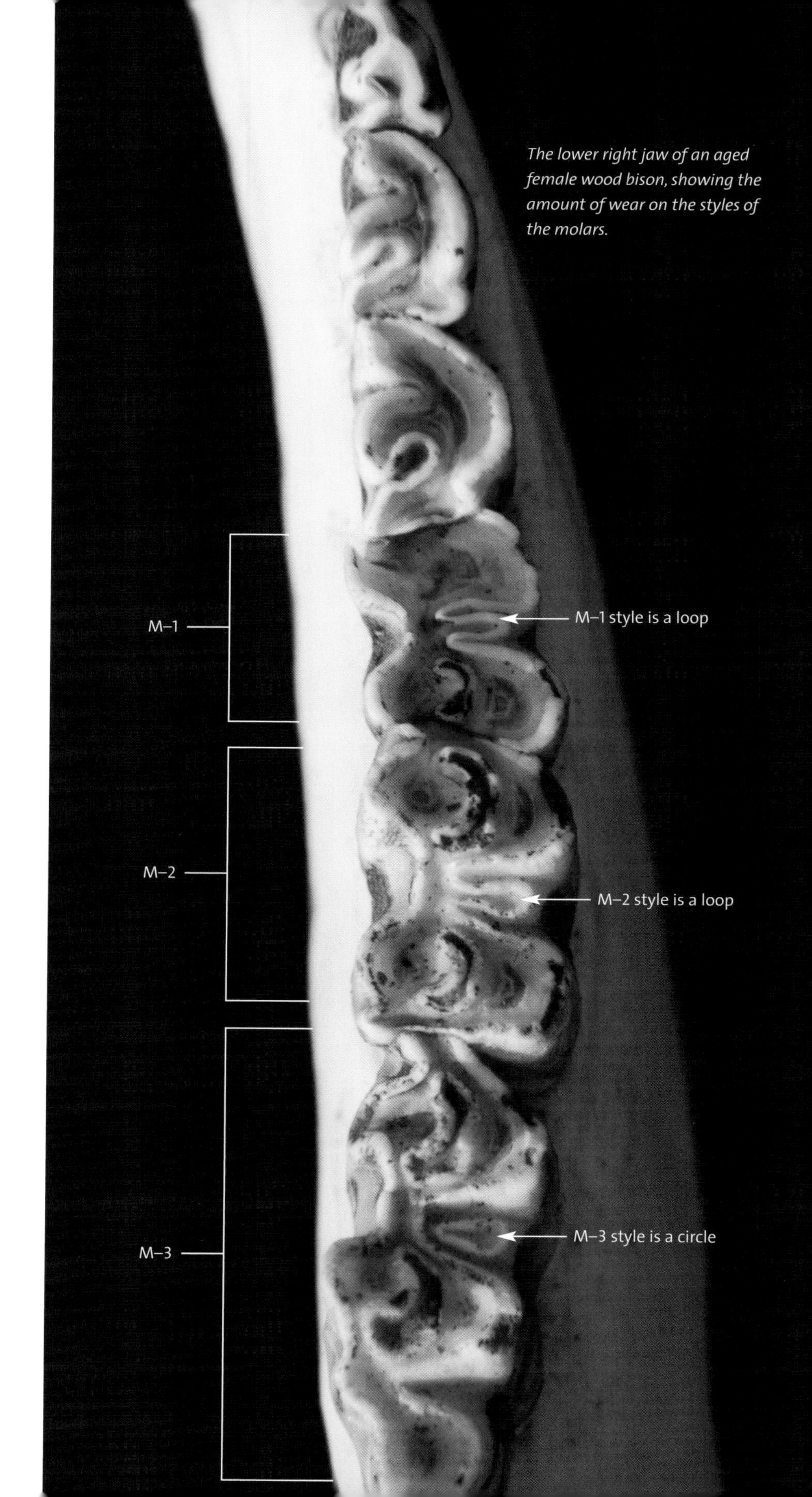

The lower right jaw of an aged female wood bison, showing the amount of wear on the styles of the molars.

Appendix 2
Public Bison Herds in North America

Table A2.1 Public Bison Herds in Canada

Note: All herds are plains bison unless otherwise noted.

Province	*Location(s)*	*Contact Information*
Alberta	Elk Island National Park *(plains & wood bison)*	Site 4, R.R. #1 Fort Saskatchewan, AB T8L 2N7 (780) 992–2950
	Hay Zama Lakes *(wood bison)*	c/o Govt. of Alberta Fish & Wildlife Division O.S. Longman Building 6909–116 Street Edmonton, AB T6H 4P2 (780) 422–9536
	Rocky Mountain House National Historic Site	Box 2130 Rocky Mountain House, AB T0M 1T0 (403) 845–2412
	Waterton Lakes National Park	Waterton Park, AB T0K 2M0 (403) 859–2447
British Columbia	Liard River *(wood bison)* Pink Mountain *(plains bison)*	c/o B.C. Dept. of Water, Lands and Air Protection 10141–101 Avenue Fort St. John, BC V1J 2B3 (604) 787–3295
Manitoba	Riding Mountain National Park	c/o General Delivery Wasagaming, MB R0J 2H0 (204) 848–7275
Northwest Territories	Nahanni River Mackenzie Bison Santuary *(wood bison)*	Box 390 Fort Smith, NWT X0E 0P0 (867) 699–3002
	Wood Buffalo National Park *(wood bison & hybrids)*	Box 750 Fort Smith, NWT X0E 0P0 (867) 872–2349
Saskatchewan	Buffalo Pound Provincial Park	#206, 110 Ominica St.West Moose Jaw, SK S6H 6V2 (306) 694–3659
	Old Man On His Back Nature Conservancy	Suite 301 1777 Victoria Avenue Regina, SK S4P 4K5 (306) 347–0447
	Prince Albert National Park	Box 100 Waskesiu Lake, SK S0J 2Y0 (306) 663–5322
Yukon	Nisling River *(wood bison)*	c/o Yukon Fish & Wildlife Box 600, Whitehorse, YT Y1A 2C6 (867) 667–5285

Table A2.2 Public Bison Herds in the United States

Note: All herds are plains bison unless otherwise noted.

State	*Location(s)*	*Contact Information*
Alaska	Chitina; Copper River Delta; Delta Junction; Farewell Lake	c/o Alaska Department of Fish & Game Box 605 Delta Junction, AK 99737 (907) 459-7236
Arizona	House Rock State Wildlife Area; Raymond State Wildlife Area	c/o Arizona Game and Fish Dept. 2222 W. Greenway Road Phoenix, AZ 85023 (928) 774-5045
California	Santa Catalina Island Conservancy	P.O. Box 2739 Avalon, CA 90704 (310) 510-1888
Colorado	Medano-Zapata Ranch	c/o The Nature Conservancy 5303 Highway 150 Mosca, CO 81146 (719) 378-2356
	Denver Parks & Recreation Herd; Daniels & Genesse Parks	c/o Denver Mountain Parks P.O. Box 1007 Morrison, CO 80465 (303) 697-4545
Illinois	Fermilab Laboratory	c/o Fermilab National Accelerator Laboratory P.O. Box 500 MS320 Batavia, Illinois 60510 (630) 840-3303
Iowa	Neal Smith National Wildlife Refuge	P.O. Box 399 Prairie City, Iowa 50228 (515) 994-3400
Kansas	Finney Game Refuge	785 South Highway 83 Garden City, Kansas 67846 (620) 276-8886
	Maxwell Wildlife Refuge	2577 Pueblo Road Canton, Kansas 67428 (620) 628-4592
	Smoky Valley Ranch; Konza Prairie Biological Station; The Nature Conservancy	Smoky Valley Ranch 1114 County Road 370 Oakley, Kansas 67748 (785) 672-3834
Kentucky	Land Between the Lakes National Recreation Area	c/o Tennessee Valley Authority 100 Van Morgan Drive Golden Pond, KY 42231 (270) 924-2061
Minnesota	Blue Mounds State Park	c/o MN Dept. of Natural Resources Route 1 Luverne, MN 56156 (507) 283-1306

State	Site	Address
Missouri	Prairie State Park	P.O. Box 97 128 NW 150th Lane Liberal, MO 64762 (417) 843-6711
Montana	National Bison Range	132 Bison Range Road Moise, Montana 59824 (406) 644-2211
Nebraska	Fort Niobrara National Wildlife Refuge	Hidden Timber Route HC14 Box 67 Valentine, NE 69201 (402) 376-3789
	Fort Robinson State Park	Box 392 Crawford, NE 69339 (308) 665-2900
	Niobrara Valley Preserve The Nature Conservancy	c/o The Nature Conservancy Route 1, Box 348 Johnstown, NE 69214 (402) 722-4440
North Dakota	Cross Ranch Nature Preserve The Nature Conservancy	c/o The Nature Conservancy 1401 River Road Center, ND 58530 (701) 794-8741
North Dakota (cont.)	Sully's Hill National Game Preserve	Box 908 Devils Lake, ND 58301 (701) 766-4272
	Theodore Roosevelt National Park	Box 7 Medora, ND 58645 (701) 623-4730
Oklahoma	Tallgrass Prairie Preserve The Nature Conservancy	c/o The Nature Conservancy P.O. Box 458 Pawhuska, OK 74056 (918) 287-4803
	Wichita Mountains National Wildlife Refuge	Route 1, Box 448 Indiahoma, OK 73552 (589) 429-3222
South Dakota	Badlands National Park	Box 6 Interior, SD 57750 (605) 433-5263
	Custer State Park	HC 83, Box 70 Custer, SD 57730 (605) 255-4515
	Ordway Prairie Reserve	c/o The Nature Conservancy 35333-115th Street Leola, SD 57456 (605) 439-3475

South Dakota (cont.)	Wind Cave National Park	R.R. #1, Box 190 Hotsprings, SD 57747 (605) 745-4600
Texas	Caprock Canyons State Park Texas State Bison Herd	P.O. Box 659 Canyon, Texas 79015 (806) 655-3782
	Clymer Meadow Nature Conservancy Preserve	P.O. Box 26 Celeste, Texas 75423 (903) 568-4139
Utah	Antelope Island State Park	4528 West 1700 South Syracuse, UT 84075 (801) 550-6165
	Henry Mountains	1084 N. Redwood Rd. Salt Lake City, UT 84116-1555 (435) 636-0262
Wisconsin	Sandhill Wildlife Area	P.O. Box 156 Babcock, WI 54413 (715) 884-2437
Wyoming	Bear River State Park	c/o Wyoming Game & Fish Dept. 5400 Bishop Blvd. Cheyenne, WY 82002 (307) 864-2176
Wyoming (cont.)	Grand Teton National Park	P.O. Drawer 170 Moose, WY 83012 (307) 739-3485
	Hot Springs State Park	c/o Wyoming Game & Fish Dept. 5400 Bishop Boulevard Cheyenne, WY 82002 (307) 864-2176
	Yellowstone National Park	c/o U.S. Dept. of the Interior National Park Service Yellowstone National Park, Wyoming 89190 (307) 344-2207

Appendix 3
A Quick Reference Guide to the Age and Sex Identification of Plains Bison

THESE PHOTOGRAPHS serve as an index to aging and sexing plains bison. When examining a bison in the field, look at it, then compare it to at the examples on the following pages. When you find one that matches, go to the page indicated to learn about how that individual fits within bison society. The codes (B0, B1, B2 , etc.) are used when recording information on the bison herd structure data sheet in Appendix 4.

Identifying Age

THE PHYSICAL APPEARANCE of bison varies considerably between one protected area and another. Indeed, within any given park or refuge there can be a wide range of body shapes, pelage characteristics, and behaviours. These variations are due in part to the slow maturation rates of bison. It takes a long time to reach old age, and during their growth such factors as nutrition, genetics, accidents, and injuries can all play a role in shaping the individual.

The bison shown here should be considered as examples of their age classes, and not necessarily definitive of them. Used as a general guide, these examples can help to assess the structure of a population, and this in turn can lead to a greater understanding of bison ecology, demographics, and life histories.

Determining Sex

WHEN WATCHING a group of bison it can be quite challenging to figure out which animal is a male and which is a female, yet for those who are responsible for caring for the herd it is essential that they be able to keep track of the number of males and females.

B0 – Male Calf

Key Characteristics

- presence of a tuft or penis sheath under belly
- horns larger than same-aged female
- ears longer than horns

Also see page 54.

B1 – Yearling Male

Key Characteristics

- well-developed tuft on penis sheath
- horns project upward at 45-degree angle
- horns longer than ears
- indistinct cape grades back along ribcage
- smaller than adult cows

Also see page 58.

B2 – Two-Year-Old Male

Key Characteristics

- horn tips point almost straight up; no inward curve
- cape line becoming distinct
- obvious penis sheath
- equal to or smaller than adult cows

Also see page 62.

B3 –Young Adult Male

Key Characteristics

- horns have begun to tilt inward at the tips
- cape demarcation sharp
- rostral swelling beginning to form
- larger than adult cows

Also see page 66.

B4 – Mature Bulls

Key Characteristics

- no ring around the upper portion of the horn
- forehead hair either fills or nearly fills the space between the horns
- firm, taut belly with no sag

Also see page 70.

B5 – Dominant Bulls

Key Characteristics

- a ring or notch around the upper portion of the horn
- horn often blunt and worn
- hair fills the space between the horns
- slight downward sag to the belly

Also see page 75.

B6 – Aged Bulls

Key Characteristics

- obvious downward sag to the belly
- all physical traits very well developed
- hair completely fills the space between the horns and may obscure them

Also see page 78.

C0 – Female Calf

Key Characteristics

- smooth underbelly
- horns smaller than those of males of the same age
- external genitalia visible under tail

Also see page 55.

C1 – Yearling Female

Key Characteristics

- thin horns equal to or longer than ears
- smooth, uninterrupted belly
- smaller than male yearlings and adult cows

Also see page 59.

C2 – Two-Year-Old Female

Key Characteristics

- thin horns curve gently upward but not inward
- flat face; poorly developed rostral swelling
- poorly developed bonnet
- roughly the same body size as a yearling male

Also see page 63.

C3 – Adult Females

Key Characteristics

- well-developed bonnet between the horns becomes larger and longer with age
- horns curve inward and backward at the tips
- wide variety of horn shapes
- udder visible on all cows that have produced a calf

Also see pages 69 and 71.

C4 – Aged Females

Key Characteristics

- well-developed rostral swelling on the bridge of the nose, giving rounded appearance
- often swaybacked
- one or both horns may be missing or misshapen
- large, full bonnet

Also see page 79.

The bison on the left is a two-year-old bull, and the other is a two-year-old female. Note the size and shape of their horns. The female's horns are more slender and delicate than those of the male.

A Quick Guide to Sex Determination

For bison older than calves, a rear view alone enables you to confirm the animal's sex. The upper photograph shows an adult male. With just the anal opening in view, the animal is obviously male. His testes are visible between is legs. The lower photograph shows an adult cow. Her vulva is visible, as is her udder.

Appendix 4
Bison Herd Structure Data Sheet

The information provided in the preceding chapters and Appendix 3 provides you with the skills necessary to record the age and sex structure of a group of bison. This data is used to capture this information. The next time you see a herd, take the time to record what you see by using tick marks (卌) and you will find the time spent watching bison far more rewarding.

	Calves			Cows				Bulls						
Date D/M/Y	B–0 Male	C–0 Female	? Unknown	C–1 Yearling	C–2 2 Yr. Old	C–3 Adult	C–4 Aged	B–1 Yearling	B–2 2 Yr. Old	B–3 Young	B–4 Mature	B–5 Dominant	B–6 Aged	Group Total
Location of Observations and Comments														
Location of Observations and Comments														
Location of Observations and Comments														
Location of Observations and Comments														
Location of Observations and Comments														

Further Reading

Allen, J.A. 1876. *The American Bisons, Living and Extinct. Memoirs of the Museum of Comparative Zoology.* Cambridge: Harvard College. New York: Arno Press, 1974.

Barsness, L. 1985. *Heads, Hides and Horns: The Complete Buffalo Book.* Fort Worth: Texas Christian University Press.

Berger, J. 1989. Female reproductive potential and its apparent evaluation by male mammals. *Journal of mammalogy* 70:347–58.

Branch, E.D. 1929. *The Hunting of the Buffalo.* New York: D. Appleton and Co.

Bryan, L. 1991. *The Buffalo People: Prehistoric Archaeology on the Canadian Plains.* Edmonton: University of Alberta Press.

Carbyn, L.N. 2003. *The Buffalo Wolf: Predators, Prey and the Politics of Nature.* Washington: Smithsonian Books.

Carl, G.R., and C.T. Robbins. 1988. The Energetic Cost of Predator Avoidance in Neonatal Ungulates: Hiding Versus Following. *Canadian Journal of Zoology* 66:239–46.

Coder, G.D. 1978. The National Movement to Preserve the American Buffalo in the United States and Canada Between 1880 and 1920. PhD diss., Ohio State University.

Conrad, L., and J. Balison. 1994. Bison Goring Injuries: Penetrating and Blunt Trauma. *Journal of Wilderness Medicine.* 5:371–81.

Coppedge, B.R., and J.H. Shaw. 2000. American Bison (*Bison bison*) Wallowing Behaviour and Wallow Formation on Tallgrass Prairie. *Acta Theriologica* 45(1):103–10.

Dary, D.A. 1989. *The Buffalo Book: The Full Saga of the American Animal.* Chicago: Swallow Press/Ohio University Press.

Daubenmire, R. 1985. The western limits of the range of the American bison. *Ecology* 66(2):622–24.

Duffield, L.F. 1973. Aging and Sexing the Post-Cranial Skeleton of Bison. *Plains Anthropologist* 18:132–39.

Egerton, P.J.M. 1962. The Cow-Calf Relationship and Rutting Behaviour in the American Bison. MSc thesis, University of Alberta.

Englehard, J.G. 1970. Behavior patterns of American Bison calves at the National Bison Range, Moise, Montana. MSc Thesis, Central Michigan University, Mt. Pleasant, Michigan.

Estes, R.D. 1976. The Significance of Breeding Synchrony in the Wildebeest. *East African Wildlife Journal* 14:135–52.

Foster, J., D. Harrison, and I.S. MacLaren, eds. 1992. *Buffalo.* Alberta Nature and Culture Series. Edmonton: University of Alberta Press.

Fryxell, F.M. 1928. The Former Range of the Bison in the Rocky Mountains. *Journal of Mammalogy* 9:129–39.

Fuller, W.A. 1959. The Horns and Teeth as Indicators of Age in Bison. *Journal of Wildlife Management* 23:342–44.

———. 2002. Canada and "Buffalo," *Bison bison*: A Tale of Two Herds. *Canadian Field Naturalist* 116:141–59.

Garretson, M.S. 1938. *The American Bison: The Story of its Extermination as a Wild Species and its Restoration Under Federal Protection.* New York: New York Zoological Society.

Geist, V. 1996. *Buffalo Nation: History and Legend of the North American Bison.* Saskatoon: Fifth House.

Geist, V., and P. Karsten. 1977. The Wood Bison (*Bison bison Athabascae* Rhoads) in Relation to Hypothesis on the Origin of the American Bison (*Bison bison* Linnaeus). *Zeitschrift fur Saugetierkunde* 42:119–27.

Giles, R.H., Jr. 1971. *Wildlife Management Techniques.* 3rd ed. Washington: The Wildlife Society.

Gordon, I.J., and A.W. Illius. 1988. Incisor Arcade Structure and Diet Selection in Ruminants. *Functional Ecology* 2:15–22.

Green, W.C.H. 1986. Age-Related Differences in Nursing Behavior among American Bison Cows. *Journal of Mammalogy* 67 (4):739–41.

———. 1992a. The Development of Independence in Bison: Pre-Weaning Spatial Relations Between Mothers and Calves. *Animal Behavior* 43:759–73.

———. 1992b. Social Influences on Contact Maintenance Interactions of Bison Mother and Calves: Group Size and Nearest-Neighbor Distance. *Animal Behavior* 43:775–85.

Green, W.C.H., and J. Berger. 1990. Maternal Investment in Sons and Daughters: Problems of Methodology. *Behavioral Ecology and Sociobiology* 27:99–102.

Green, W.C.H., and A. Rothstein. 1991. Sex Bias or Equal Opportunity? Patterns of Maternal Investment in Bison. *Behavioral Ecology and Sociobiology* 29:373–84.

———. 1991. Tradeoffs Between Growth and Reproduction in Female Bison. *Oecologia* 86:521–27.

———. 1993a. Asynchronous Parturition in Bison: Implications for the Hider-Follower Dichotomy. *Journal of Mammalogy* 74 (4):920–25.

———. 1993b. Persistent Influences of Birth Date on Dominance, Growth and Reproductive Success in Bison. *Journal of Zoology* 230:177–86.

Green, W.C.H., J.G. Griswald, and A. Rothstein. 1989. Post-Weaning Associations among Bison Mothers and Daughters. *Animal Behavior* 38:847–58.

Guthrie, R.D. 1990. *Frozen Fauna of the Mammoth Steppe: The Story of Blue Babe*. Chicago: University of Chicago Press.

Halloran, A.F. 1961. American Bison Weights and Measurements from the Wichita Mountains Wildlife Refuge. *Procedures of the Oklahoma Academy of Science* 41:212–18.

Hart, R.H. 2001. Where the buffalo roamed—or did they? *Great Plains Research* 11 (Spring 2001):83–102.

Haynes, G. 1984. Tooth wear rate in northern bison. *Journal of Mammalogy* 65(3):487–91.

Helbig, L. 2005. Onset of puberty and seasonal fertility in bison (*Bison bison*) bulls. MSc thesis, University of Saskatchewan. 121 pp.

Herrig, D.M., and A.O. Haugen. 1969. Bull Bison Behavior Traits. *Iowa Academy of Science* 76:245–62.

Hewitt, C.G. 1921. The buffalo or bison: Its present, past, and future. In C.G. Hewitt, editor, *The conservation of the wildlife of Canada*. New York: Charles Scribner's Sons. Pp. 113–42.

Hornaday, W.T. 1904. The American Buffalo. Chapter 8 in *The American Natural History*. New York: Charles Scribner's Sons. Pp. 99–103.

Hudson, R.J., and S. Frank. 1987. Foraging Ecology of Bison in Aspen Boreal Habitats. *Journal of Range Management* 40 (1):71–75.

Ims, R.A. 1990a. On the adaptive value of reproductive synchrony as a predator swamping strategy. *Am. Nat.* 136:485–98.

Komers, P.E., F. Messier, and C.C. Gates. 1993. Group Structure in Wood Bison: Nutritional and Reproductive Determinants. *Canadian Journal of Zoology* 71:367–71.

Komers, P.E., K. Roth, and R. Zimmerli. 1992. Interpreting Social Behaviour of Wood Bison Using Tail Postures. *Zeitschrift fur Saugetierkunde* 57:343–50.

Lott, D. 1979. Dominance Relations and Breeding Rate in Mature Male American Bison. *Z. Tierpsychol.* 49:418–32.

———. 1981. Sexual Behaviour and Inter-Sexual Strategies in American Bison. *Z. Tierpsychol* 56:97–114.

———. 1991. American Bison Socio-ecology. *Applied Animal Behavioural Science* 29:135–45.

Lott, D.F., and J.C. Galland. 1985. Parturition in American Bison: Precocity and Systematic Variation in Cow Isolation. *Z. Tierpsychol* 69:66–71.

———. 1987. Body Mass as a Factor Influencing Dominance Status in American Bison Cows. *Journal of Mammalogy* 68 (3):683–85.

Lott, D., and S.C. Minta. 1983. Random Individual Association and Social Instability in American Bison. *Z. Tierpsychol* 61:153–72.

MacEwan, G. 1995. *Buffalo: Sacred and Sacrificed*. Edmonton: Alberta Sport, Recreation, Parks & Wildlife Foundation.

MacLaren, I.S. 1992. Buffalo in word and image: From European origins to the art of Clarence Tillenius. In Foster, J., D. Harrison and I.S. MacLaren. *Alberta: Studies in the arts and sciences* 3(1):79–129.

Maher, C.R., and J.A. Byers. 1987. Age-Related Changes in Reproductive Effort of Male Bison. *Behavioral Ecology and Sociobiology* 21:91–96.

McDonald, J.N. 1981. *North American Bison: Their Classification and Evolution*. Berkeley: University of California Press.

McGinley, M.A. 1984. The Adaptive Value of Male-Biased Sex Ratios Among Stressed Mammals. *American Naturalist* 124:597–99.

McHugh, T. 1958. *The Time of the Buffalo*. Lincoln: University of Nebraska Press.

Meagher, M. 1973. The Bison of Yellowstone National Park. *National Park Service Scientific Monograph Series* 1:1–161.

Melton, D.A., N.C. Larter, C.C. Gates, and J.A. Virgl. 1989. The Influence of Rut and Environmental Factors on the Behaviour of Wood Bison. *Acta Theriologica* 34 (12):179–93.

Moodie, D.W., and A.J. Ray. 1976. Buffalo Migration on the Canadian Plains. *Plains Anthropologist* 21:45–52.

Myers, J.H. 1978. Sex Ratio Adjustments under Food Stress: Maximization of Quality or Numbers of Offspring? *American Naturalist* 112:381–88.

Novakowski, N.S. 1957. Aerial re-survey of bison in Wood Buffalo National Park and surrounding areas, 1957. Unpublished report CWSC-216. Canadian Wildlife Service (Edmonton, AB). 12 pp.

Post, D.M., T.S. Armbrust, E.A. Horne, and J.R. Goheen. 2001. Sexual segregation results in differences in content and quality of bison (*Bos bison*) diets. *Journal of Mammalogy* 82 (2):407–13.

Reynolds, H.W., C.C. Gates, and R.D. Glaholt. 2003. "Bison (*Bison bison*)." In G. A. Feldhamer, B. Thompson, and J.A. Chapman, eds. *Wild Mammals of North America: Management and Conservation*. 2nd ed. Baltimore: Johns Hopkins University Press.

Rhoads, S.N. 1897. Notes on living and extinct species of North American Bovidae. *Proceedings of the Academy of Natural Sciences* 49:483–502.

Roe, F.G. 1970. *The North American Buffalo: A Critical Study of the Species in its Wild State*. 2nd ed. Toronto: University of Toronto Press.

Rothstein, A., and J.G. Griswold. 1991. Age and Sex Preferences for Social Partners by Juvenile Bison Bulls. *Animal Behavior* 41:227–37.

Rutberg, A.T. 1983. Factors Influencing Dominance Status in American Bison Cows. *Z. Tierpsychol* 63:206–12.

———. 1984. Birth Synchrony in American Bison: Response to Predation or Season? *Journal of Mammalogy* 65 (3):418–23.

———. 1986. Lactation and Fetal Sex Ratios in American Bison. *American Naturalist* 127:89–94.

———. 1987. Adaptive Hypothesis of Birth Synchrony in Ruminants: An Interspecific Test. *American Naturalist* 130:692–710.

Schorger, A.W. 1945. The bison in Florida. *Journal of Mammalogy* 26(4):432–33.

Schult, M.J. 1979. *Where Buffalo Roam*. Interior, SD: Badlands Natural History Association.

Shaw, J.H., and T.S. Carter. 1988. Long-Term Associations Between Bison Cows and Their Offspring: Implications for the Management of Closed Gene Pools. In *Procedures of the North American Bison Workshop,* ed. J. Malcomb. Missoula: n.p.

———. 1989. Calving Patterns among American Bison. *Journal of Wildlife Management* 53 (4):896–98.

Stelfox, J. Brad, ed. 1992. Alberta's Hoofed Mammals: Their Ecology, Status and Management. Edmonton: n.p.

Taylor, D. 1990. *The Bison and the Great Plains*. Animals and Their Ecosystems Series. Toronto: Crabtree Publishing.

van Zyll de Jong, C.G. 1986. A Systematic Study of Recent Bison, with Particular Consideration of the Wood Bison (*Bison bison athabascae*, Rhoads 1898). National Museum of Natural Sciences, *Publications in Natural Sciences* 6.

van Zyll de Jong, C.G., C. Gates, H. Reynolds, and W.E. Olson. 1995. Phenotypic Variation in Remnant Populations of North American Bison. *Journal of Mammalogy* 76 (2):391–405.

van Zyll de Jong, C.G., C.C. Gates, W.E. Olson, H. W. Reynolds, and T. Unka. 1991. *An Illustrated Guide to Bison Phenotypes*. N.p.: Taxonomy Subcommittee of the Wood Bison Recovery Team.

Wilson, G.A., and K. Zittlau. 2003. COSEWIC Status Report on the Plains Bison (*Bison bison bison*). Ottawa: Committee on the Status of Endangered Wildlife in Canada.

Wilson, G.A., W.E. Olson, and C. Strobeck. 2002. Reproductive Success in Wood Bison (*Bison bison athabascae*), Established Using Molecular Techniques. *Canadian Journal of Zoology* 80:1537–48.

Wolff, J.O. 1998. Breeding Strategies, Mate Choice, and Reproductive Success in American Bison. *Oikos* 83:529–44.

About the Author and the Photographer

WES OLSON was born and raised in the foothills of western Alberta, and has worked with wildlife most of his life. Since the early 1980s, he has spent his career as a park warden at Elk Island National Park in central Alberta. There he became fascinated with bison. This book is the culmination of more than twenty years of observing bison behaviour. His drawings clearly depict this magnificent species and make understanding their behaviour much easier.

JOHANE JANELLE migrated west from Quebec in 1982 to explore the mountains on horseback and thus began a life-long passion for wild places and the animals that live in them. She has travelled extensively with her husband, Wes Olson, to document the plains and wood bison of North America. Today she runs a successful custom equine photography business. Her images of the horse can be viewed regularly on the covers of a variety of horse magazines in Canada and the United States, and on her website, www.johanejanelle.com.

Wes and Johane own and operate the Broken Lantern Ranch, where they raise purebred plains bison and Canadian warmblood horses.